ADVANCE PRAISE

Heart-Centered Marketing is an insightful, informative, and authentic book with an abundance of stories that will resonate with female entrepreneurs worldwide. The style of this anthology is warm and inclusive, with an easy-to-follow flow and a friendly tone. Each chapter is shared from the heart and covers everything from the foundations of business through to shifting your mindset and embracing specialist skills such as speaking engagements and public relations. This book is a must-have resource for any entrepreneur who wants to understand the value behind support, connection, and nurturing yourself to build a thriving business.

— SHELLEY WILSON, AUTHOR AND WRITING MENTOR

Wow! What an impactful, insightful, and empowering book. This is a must read for those who are beginning their entrepreneurial journey, as well as for those who are "seasoned" entrepreneurs. This anthology has left no stone unturned when it comes to the ins and outs of marketing, attracting clients, and nurturing them. Providing readers with a wealth of knowledge, this is your one-stop shop, so to speak. There are takeaways for everyone within the pages of this book—a truly phenomenal collaboration of marketing strategies! This is what women empowering others looks like!

— KIM BARK WHITE, OWNER OF MINDFUL EMPOWERMENT

This is a great book for beginning entrepreneurs. It gives a good overview of many of the problems that entrepreneurs come up against when they are starting out—or even just thinking about starting out. If you find one of the topics to be interesting, you can get a good introduction to the basics in that chapter, and then know exactly where you want to delve deeper in your research. A good reference to have to hand!

—JULIA POGER, KNOW YOUR WORTH

What a wonderful tool of wisdom and inspiration for all women entrepreneurs, written by successful women entrepreneurs! As an entrepreneur, I connected with the challenges, rewards, and experiences highlighted by the authors. This book offers encouragement and tips to keep in mind in many aspects of entrepreneurship including overcoming negative thinking, discovering your value and self-worth, public speaking, and attracting your ideal clients. Whether you are currently considering entrepreneurship or are a seasoned business owner, the excerpts in *Heart-Centered Marketing: Proven Strategies That Naturally Attract and Nurture Clients* offers sound guidance to help you achieve success along your journey.

— GINA RAMSEY, HUMORIST AND AUTHOR OF *BURNT GLOVEBOXES: EMBRACING LIFE WHEN IT GOES UP IN FLAMES*

The anthology, *Heart-Centered Marketing: Proven Strategies That Naturally Attract and Nurture Clients,* has an all-star line-up of seasoned businesswomen sharing their wisdom about all things marketing for small business owners. I personally know about half of them and can vouch for their expertise in their fields! Every chapter includes sage insights and immediately actionable advice for growing your business. After you read this book, you'll be able to start attracting and working with clients you actually love and watch your business bloom!

— CHERI D. ANDREWS, ESQ., AUTHOR OF
SMOOTH SAILING: A PRACTICAL GUIDE
TO LEGALLY PROTECTING YOUR BUSINESS

Heart-Centered Marketing: Proven Strategies That Naturally Attract and Nurture Clients is not just a book; it's a revolution in the world of marketing. As a seasoned entrepreneur with a rich history of experience, I approached this anthology with a blend of skepticism and curiosity. The concept of heart-centered marketing, a term that seemed almost oxymoronic in the cutthroat world of business, intrigued me. What I discovered within its pages was not only a reaffirmation of my belief in authenticity but also a treasure trove of practical insights that reshaped my marketing approach.

— SUZANNE TAYLOR-KING MCPC, CHLC,
CAPP, BUSINESS COACH,
EUDAIMONOLOGIST, AUTHOR, SPEAKER,
AND CHANGE AGENT

Heart-Centered Marketing: Proven Strategies That Naturally Attract and Nurture Clients is a must-read for anyone who has dreamed of starting a business – or who has already started and wonders what they're doing wrong! Each chapter has insight to identify where a business owner can go awry – I saw myself in several chapters, in fact! More valuable, though, are the action steps each author offers that a business owner can take to right their ship and move forward to grow an impactful, profitable business. Easy to read, either front-to-back or a chapter-at-a-time based on what calls to you. Worth the read!

— **CLARISSA CONSTANTINE, FOUNDER OF THE PARENT(W)EEN CONNECTION**

The simplest path to success as an entrepreneur is finding a version of marketing that actually fits YOU, and *Heart-Centered Marketing* is a must-read resource for doing exactly that! Each and every chapter of this book takes lessons from the full spectrum of women entrepreneurs and turns them directly into gold (both in nugget and whole-bar form!) for the reader! If you're a woman entrepreneur, *Heart-Centered Marketing* is a resource you'll go back to time and time again as you discover YOUR way to effectively and successfully market what you have to give the world, heart-first.

— **SARA TORPEY, BUSINESS COACH**

Heart-Centered Marketing provides straightforward, simple, and SAGE advice to women as they begin the process of building a business. As a bodyworker and healer in private practice for the last thirty years, I can attest to the noble truths spoken in the pages: Trust your self-worth, learn how to attract the types of clients you want to serve, and celebrate connections with other entrepreneurs in order to build community and reciprocal support. Collaboration and trust will bring more joy and abundance to whatever you put your mind to. Building a business with heart is so rewarding!

— AMY BOND TALIAFERRO, LMT, CSP, SEP, OWNER OF INTEGRATIVE BODYWORK

What an amazing resource! *Heart-Centered Marketing* speaks to the giver, helper, and teacher in me. As a heart-centered coach for five years, I have often found it difficult to sort through and locate advice that resonates with me. This book has it all in one place. The topics—mindset, fundamentals, visibility and sales—are shared from the heart with relatable stories and practical tips. It's a book I'll come back to again and again.

— BARB HUBBARD, TIME MANAGEMENT COACH & ADHD SPECIALIST

Having the confidence to share our knowledge and skills with our patients and clients is a barrier we must break through. In reading *Heart-Centered Marketing*, I felt as if the authors for each chapter were speaking to me and addressing my fears and hesitation. This is a brilliant collection of experiences, tools, and learnings, which emphasize the importance of celebrating ourselves and our accomplishments. I found reading *Heart-Centered Marketing* can help guide one's journey as an entrepreneur as well as in life in general.

— DR. RADHIKA CHAWLA, O.D., OWNER AND PRESIDENT OF RICHMOND HILL OPTOMETRIC CLINIC

I loved reading each author's experience, so much of which resonated with my own. When you start out as an entrepreneur, mindset is everything, and I love that that is where the book begins. In fact, mindset was a constant throughout the book, just like it is in our experience. Each chapter provides lived experiences, unique insights, and takeaways that are actionable and will move you forward. This is a great read and also a great reference point as I continue to grow my business.

— KIRSTEEN WILLIAMSON-GUINN, ELEVATE WOMEN

I have an alternative healing practice, and I've found that many of the traditional ways of marketing don't fit with a spiritual business and—more importantly—nor do they fit with the clients that would be drawn to my services. More warmth and a heart-to-heart connection are needed. *Heart-Centered Marketing* contains nineteen chapters, each with their own lesson on how to be authentic in marketing. This is a great resource for anyone running a business who wants something other than the common cookie-cutter approach.

— DEBRA MORRILL, SHAMANIC HEALER AND TEACHER

If only this book had been around when I quit corporate life. That role felt masculine and rigid and I wanted to do it differently. However, I soon realized that having skills and an idea does not make it easy to set up and run a business. I longed for a manual and reassurance that self-doubt is normal. This compilation of highly relatable experiences, mentorship, and wisdom from other women business owners does the job and is a good reminder that feminine power is far from fluffy.

— TRACY SHORT, TRACY SHORT & CO

Heart-Centered Marketing is a must-read for professionals looking to enhance their marketing strategies. As professionals, we often get caught up in the numbers and forget about the human element of marketing. This book reminds us to put people first and build meaningful client relationships. I highly recommend *Heart-Centered Marketing* for anyone looking to take their marketing efforts to the next level.

— MINDY VAN VLEET, REGIONAL DIRECTOR, POLKA DOT POWERHOUSE

Heart Centered Marketing is the book I wish I'd had when I started my business. It's practical, full of brilliant advice, and one to keep coming back to time and time again. In distilling the lived experience of many talented women into easily digestible learnings, it provides a simplicity and ease that gives a refreshing alternative to the hustle culture that's so often preached. It's a treasure trove of insights, guidance, and reassurance that I'm not alone in wanting to realize my ambition without sacrificing my values.

—JACQUI JAGGER, THE PRACTICAL
LEADERSHIP COACH

This is a MUST HAVE book for any entrepreneur and business owner. The information and tips each author shares are easy to implement. I learned new ways of marketing that don't require a lot of time, yet will lead to my business having a larger, more impactful presence. I will reread this book often to be certain I'm not leaving important steps out. Go get this book today!

—MARGARET MARTIN, RE-DESIGN YOUR
LIFE COACHING AND AUTHOR OF
*THE CHATTER THAT MATTERS:
YOUR WORDS ARE YOUR POWER*

Step into an era of marketing where success isn't about hard sells but genuine connection. *Heart-Centered Marketing* is your passport to a thriving business with heart. Settle in and immerse yourself in this anthology of industry visionaries, where feminine energy transforms marketing into an art of attraction, not pursuit. Discover the secrets of skyrocketing self-worth, magnetic branding, and empathetic communication. Embrace this game-changing guide today and experience how the power of heart can elevate your business.

— REGINA GARAY, FOUNDER OF RIGHT BRAIN LOVES LEFT

Heart-Centered Marketing so hugely inspires me. The women who have contributed their time and brilliance are inspirational, each in their own right. I'm grateful to Deborah Kevin, MA, and Jill Celeste, MA, for creating the idea of this book and bringing it to fruition so that women entrepreneurs everywhere can benefit from its teachings. I love that this book can be read from cover to cover and be a resource for whatever specific area of challenge in the moment. This book deserves a very special place on the bookshelves of women entrepreneurs everywhere!

— KRISTY L. CRIPPEN, FOUNDER OF COLLABORATIVE PARTNERS ADMINISTRATIVE SERVICES, LLC

Heart-Centered Marketing is a powerful collection of passionate and personal essays by successful entrepreneurs and business operators. If you are looking for the optimal way to drive your business forward and reach your full potential, whilst staying true to your feminine nature and embracing your worth, then look no further. With clarity, humor, and a unique blend of wisdom and storytelling, *Heart-Centered Marketing* will open the door to the future you have been waiting for.

— **NORA TANDBERG, CHIEF FINANCIAL OFFICER AT PAPIRFLY**

We remember key learning points best when they are driven home with a relatable story. This collection is filled with heartwarming, insightful and occasionally hilarious vignettes to remind us what matters most in marketing—our connections with ourselves, our clients, our values, and our purpose. Thought leaders seeking more clarity around their goals and dreams will glean helpful insights in this well organized, easy to read compilation. For busy entrepreneurs, bite-sized reading and learning is best and this book fits the bill!

— **ELIZABETH BARBOUR, AUTHOR OF *SACRED CELEBRATIONS: DESIGNING RITUALS TO NAVIGATE LIFE'S MILESTONE TRANSITIONS***

Heart-Centered Marketing is a must-read book for all women who think marketing yourself and your business is icky. There is another way that shares you with the people that need you, people you will love to work with, and people that will love working with you: a heart-centered way, where the people you touch will become part of your marketing team by singing your praises.

— **CONNIE JO MILLER, ENIGMA BOOKKEEPING SOLUTIONS**

HEART-CENTERED MARKETING

HEART-CENTERED MARKETING

PROVEN STRATEGIES THAT NATURALLY ATTRACT AND NURTURE CLIENTS

DEBORAH KEVIN, MA

JILL CELESTE, MA

HIGHLANDER
PRESS

ISBN: 978-1-956442-22-9
Ebook ISBN: 978-1-956442-23-6
Library of Congress Control Number: Applied for.

Published by Highlander Press
501 W. University Pkwy, Ste. B2
Baltimore, MD 21210

Cover design: Hanne Brøter
Managing Editor: Deborah Kevin, MA
Associate Editor: Jill Celeste, MA

For all those entrepreneurs who have longed to tap fully into their feminine energy and naturally attract and nurture ideal clients: consider this your permission slip.

CONTENTS

INTRODUCTION

DEBORAH KEVIN, MA

In 2023, Highlander Press published its first anthology, *Your First Year: What I Wish I'd Known*, to rave reviews. It made a difference for our intended audience: women entrepreneurs who were either new in their businesses or considering launching one. That book was chock full of practical wisdom, and I naturally thought, "What could we contribute next in service to women entrepreneurs?"

That question led to me research what books were out there written for women entrepreneurs by women entrepreneurs. I came to an astonishing discovery: despite women owning over twelve million businesses employing over ten million people[1], only three of the top forty marketing books on Amazon were written by women.

Marketing theory and practice has long consisted of cold calling, business card exchanges, hard sales tactics, aggressive funnels, and the like. We wanted to highlight and leverage feminine energy instead. Attract versus pursue. Nurture instead of land. Build relationships. Foster connection. Support and encourage. Hence the title and message of this book: you can attract and nurture your ideal clients AND be successful in business!

The authors in this book span a variety of industries and have various lengths of time as business owners; we've made this mindful decision to cover the gambit of experiences and hope that you find exactly what you need right now—and refer often to the gifts included.

To make the most of this powerful resource, you might scan the table of contents and land on a topic you're called to. Or you could read this book cover to cover. I hope you dog-ear the pages, highlight significant passages, and draw on inspiration that takes you to the next level. Connect with the authors on LinkedIn or social media. Together, we can make a positive impact on their world!

We have your back and want you to be wildly successful! Our world desperately needs more courageous women like you.

Let's do this—together!

Deborah Kevin, MA
Chief Inspiration Officer
Highlander Press

MINDSET

1

SHIFTING THOUGHTS ABOUT MY SELF-WORTH

FELICIA MESSINA-D'HAITI

After I completed my official feng shui and space-clearing certification more than fifteen years ago, I was so excited to work with my first client. The excitement was a blend of joyful anticipation mixed with some nervousness about taking on the first paid client. Even though I had done amazing things during my training and with my practice clients, I was a bit unsure about taking the next step. Starting on a new pathway also felt risky, as I have a job that I enjoyed (most of the time) and was really good at doing. Yet, I felt that something else was tugging at my heart, and I wanted to explore this new desire. When I was teaching, I became aware of how the classroom space impacted the atmosphere and the learning within, so I stepped into formally learning feng shui.

Shortly after I earned my certification, a friend called to say that she referred me to another entrepreneur who wanted a space clearing for her new business. I connected with this interior designer and explained what services I provide and the cost for services. However, I was not prepared for how she grilled me about my training, past experience, and more. Thankfully, she agreed to the price, and we set a date for the space clearing. I began to plan for my first paying client.

I was so elated about the opportunity to use the skills that I had been practicing.

Several days later, I received an unpleasant email from her, stating she had much more experience than me, that my price was too high based on my experience, and that she can do the space clearing herself and do a better job. I was so hurt and disappointed.

What I did not realize at the moment was that a tiny seed of self-doubt that had been buried in my being had taken root and began to grow more quickly. It's true that I did not have a great deal of experience; however, that did not correlate to my effectiveness. While I tried to ignore what was stirred up, it persisted.

Not long after, I joined a networking group for women entrepreneurs where I was speaking with another feng shui and space-clearing consultant. Through the seemingly pleasant conversation, she continually emphasized the years of training and apprenticeship that she underwent to earn her certification, implying that I had not yet done enough work to offer services to clients. My seed of self-doubt continued to grow as I began to compare myself to others in the same field. At the time, I did not understand the impact these thoughts had on my journey.

Even when I was given thank-yous and compliments on my services and the courses I taught, the seed continued to grow. I felt that nothing I did was good enough. I became an excellent detective in identifying self-criticisms and imperfections to add to the growth of this seed. And it was often so subtle that I did not even realize what was happening.

I started to resist things like promoting myself, setting prices, taking risks, and venturing out to explore new areas. I supported these thoughts through procrastination and complete deferral. It became difficult to graciously receive compliments without downplaying my accomplishments or pointing out what didn't go right. But I managed to move forward in spite of myself.

Years later, I signed up for a course that focused on self-development. In one of the classes, I participated in a process to identify an unsupportive core belief. The unsupportive core belief that came up for me was "not worthy." How about that?! There it was right on the screen in front of me. This identification process was a gift for me because it forced me to identify it, call it out, and deal with it.

Through further exploration, I became aware that this belief went much farther back than my interactions with the two entrepreneurs. I could see evidence of it in the ways that I chose to "prove myself" while growing up and especially in the workplace. I saw it in other instances where others judged me as not doing "enough" to reach different milestones. Instead of brushing it off as *their* insecurity, I took it on as mine.

For example, I became a teacher through a nontraditional certification route, already with a wealth of experience in the arts and museum field. Yet, I was criticized specifically for not having a traditional teaching degree. My reaction was to obtain a national teaching certificate, becoming only the third arts teacher in the entire school district with that certification. And while that was an amazing experience and accomplishment, I did not pause to celebrate myself or recognize what I had done. I felt it had to be done to demonstrate my worthiness.

After I became aware of the depth of the beliefs and feelings that I had supported about my own self-worth, I made a conscious effort to shift my mindset. I made an intentional effort to shift my thinking away from the self-criticism, comparison, and taking on the negative things people may say.

Reframing is one tool that I have used to shift my thoughts. I noticed when I created a self-criticizing thought, and I would reframe that thought. I began to see myself as my own first client and would coach myself through these thoughts and feelings—seeking out the root, forgiving myself for thinking about or talking to myself in less than positive ways, and finding new ways to talk to and about myself.

Another supportive strategy I began to use was to celebrate my accomplishments, both big and small. I noticed that my lack of self-worth resulted in not celebrating my accomplishments. Things became a checklist to finish in order to demonstrate worthiness. Yet, this behavior just exacerbated the issue. Quite similar to a gratitude list, I began to journal daily in celebration of what I had accomplished that day, no matter how big or small the experience. I paused between projects to reflect what had accomplished. Even in the cycles of nature, there is a pause between the harvest and the planting of new seeds in the spring. This pause includes celebration, rest, and reflection.

Another way to reset my self-worth was to practice gratitude about my life's journey. There is joy and a wealth of experience in not walking on a straight path. Instead of downplaying my seemingly unrelated experiences, I embraced the variety of experiences, and looking for and creating points of intersection in my journey. Along with celebrating my accomplishments, I intentionally practice gratitude for all that has occurred on my life's journey, including all of the wonderful things, but also the challenges, grief, and sadness that has occurred. Everything has contributed to the whole picture of who I am, which is definitely not the same as anyone else.

A huge and challenging shift for me was releasing what others think about me, including saying no to people and things that I did not feel aligned with. Before saying yes, I pause and think about the time and energy involved, and the alignment of the "ask" with my values and other life choices. When I start to wonder about what others are thinking about me or my "no," I remind myself that what they are thinking is their experience, not mine. I refocus on my experience, and I honor what my spirit needs to feel loved and joyful.

Incorporating these shifts into my daily practice have created a new outlook about myself and the work I do in this world. By reframing negative thoughts, celebrating accomplishments, practicing gratitude, and releasing what others think about me, I've been able to alter how I feel about myself and the experiences I have each day. I listen to myself more and care for myself more.

At the root of embracing self-worth is loving myself more, as I am, and appreciating my life's journey, with all its twists and turns. While

I delve more into self-development and my thoughts and feelings about myself each day, I have seen how it transforms the experiences that I have with everyone I encounter, including my clients and students.

KEY TAKEAWAYS:

- Improving your self-worth positively affects all aspects of your entrepreneurial journey, from pricing to how you work with your clients.
- If you are resisting promoting your business, or raising your prices, or exploring new ideas to grow your business, you may be experiencing low self-worth.
- Celebrate your accomplishments—no matter how big or small they are. This will help increase your self-worth.
- Also be sure to express gratitude for everything in your life and business.
- Remember that someone's opinion of you is none of your business.

2

SPEAK SUCCESS INTO BEING: THE POWER OF AFFIRMATIONS

GAIL DIXON

Let's talk about something that might sound a bit out there but has seriously transformed my entrepreneurial game—affirmations. I don't mean magical incantations or secret spells. I'm talking about the power of words to shape our reality. As an entrepreneur, I've discovered that "speaking success into being" through affirmations is like having a superpower in my back pocket.

My Story

Striding confidently to the podium, I prepared to do something I'd done hundreds of times before—deliver a speech. It should've been old hat, the easiest thing in the world. Instead, it was a disaster, one that changed my life and taught me how to use affirmations to speak things into being.

Head up, shoulders back. Gazing over the heads of people in the room, I focused on a spot in the back, I aimed my voice at the person standing against the far wall. Taking a breath, I opened my mouth to speak.

Nothing came out. Not one single word. I tried again. And a

third time. Still nothing. By now, the audience was growing restless, and I was growing more and more breathless. Signaling to my colleague seated beside the podium, I wrote a note, "Can't breathe. Can't talk." I slid my notes to her and walked away from the podium and off the platform, barely stepping off the last step before passing out.

After more tests than I'd had in all my life combined, the verdict was to send me to another hospital for yet another round of tests and specialists. Meanwhile, I couldn't breathe without the oxygen. "Paralyzed diaphragm" was the diagnosis. My life was turned upside down. I couldn't do my job. I couldn't breathe without being tethered to an oxygen concentrator or portable tank. I lived in fear of a power failure, and time away from home was measured by how much oxygen remained in the portable tank.

I felt disheartened, discouraged, and defeated. None of the doctors could determine a cause for the paralyzed diaphragm. Without a cause, they couldn't offer a prognosis, so I had no idea if my condition was as good as it would be or as bad as it would ever be. Should I hope for something better or hope for nothing worse?

Simply put, I was living a life that was totally unfamiliar to me. I didn't know who I was anymore. I thought I knew about handling change, but nothing prepared me for this loss of identity. My days were filled with those tormenting whispers: "Are you strong enough? Can you really pull this off?" I was juggling countless responsibilities, second-guessing every decision, and battling an avalanche of what-ifs. Finding my way around and through to the other side was a long and arduous process, one I navigated with help from every corner of my life.

The biggest help came from my very wise therapist, Gloria. One day she said, "Gail, you are grieving everything you think you've lost. What do you still have that can help you through?" It took some time, but I finally realized I still had the power of my gift with words. The light dawned. Maybe I'd feel differently about my situation if I spoke about it differently—if I used my words not just to describe my reality but to define it. That's when I became seriously committed to daily affirmations.

Why Affirmations?

Why affirmations? Because words matter, period. Our words express our thoughts. Our thoughts shape our beliefs, our beliefs shape our actions, and our actions shape our outcomes. So, why not use this linguistic domino effect to my advantage? What if I could change my self-talk to create a more positive reality? At Gloria's suggestion, I committed to a regular, consistent practice of gratitude and affirmations. The idea is to flip the script on the doubts, fears, and insecurities that like to creep in and replace them with uplifting, motivating self-talk.

I confess that both gratitude and affirmation were in short supply when I was at my lowest. Gratitude came more quickly than the affirmations did. Despite believing in the power of words, I had some serious reservations. First, I struggled with the idea that the affirmations didn't feel authentic. I wasn't ready to take them as true. Those positive things were not true in the moment, but they were true in the reality I was working to create. The direction to frame the affirmations from the desired future, rather than from the feeling of the moment, made all the difference.

I also had to work to silence my inner critic. She had become even more vocal than usual amid the changes I was experiencing. Every day, something that had been easy previously was now difficult, if not impossible. Questions about confidence and competence ran through my mind at regular intervals. The judgments from my inner critic also made it difficult to respond to the external voices that questioned whether I should just accept reality and give up my entrepreneurial ambitions.

Before I was willing to embrace the affirmation habit, I had to quiet my mother's remembered voice telling me not to brag or talk too highly about myself. Humility was a highly prized virtue in my family, even if it meant holding yourself back or playing small.

Finally, I had to assure myself that using affirmations was not abdicating my responsibility to work hard and take action to meet my goals and achieve success. I wasn't looking for a magic wand; I was using a tool. In fact, I was using the tool that was exactly the one I should be using because it was ideally suited to my gifts and talents.

Once I overcame my initial resistance, I was ready to create my personal affirmations. I wanted to make the affirmations unique to me, not just something I'd borrowed from some guru or influencer. I realized that for me, it was important to distinguish affirmations from encouragement or applause. Both those things come from outside, from observation, from others. They are bestowed and can be taken away at will and therefore, don't really belong to me. Affirmations, on the other hand, come from inside me as a personal declaration of identity and reality. They are entirely within my control and available to me at any time.

I began by identifying my innermost fears and negative beliefs. It was like unearthing buried treasure—only this treasure was made of self-limiting thoughts. "You're not healthy enough to lead this business," whispered one. "You're nothing special, especially now that you can't even travel to give a speech," taunted another. I became aware that my self-limiting thoughts were all rooted in the belief that I was not enough. I faced these limiting beliefs head on and transformed them into powerful affirmations.

"I am strong and capable enough to be successful," I declared to myself. "My ideas are valuable and will inspire positive change," was another. These affirmations became my personal mantras, my shield against self-doubt. Thinking back to the common root of my self-doubts, not being enough, I created this affirmation series as part of my daily morning practice:

- I am strong enough.
- I am capable enough.

- I am enough.

Every morning, after I express my gratitude aloud, I recite these affirmations. It felt a little awkward at first, but over time, with each word, a spark of something magical ignited within me. These were no longer mere words; they were my guiding stars. As the days turned into weeks, the affirmations became less like lines to recite and more like the rhythm of my heartbeat.

I was thrilled when my affirmation practice started to yield results. It wasn't overnight, but it was undeniable. I noticed that my once-paralyzing doubts were replaced with a newfound confidence. The great ideas were translated into action more completely and consistently. Conversations with potential collaborators flowed effortlessly, as if the universe itself was paving the way. It was like affirmations were turning my thoughts into actions, and my actions into a cascade of wins.

Affirmations: Not Magic, but Powerful

So, here's the deal: Affirmations aren't some mystical incantations that instantly manifest miracles. They're not a substitute for hard work, strategy, or resilience. But they are a potent tool that rewires your mindset, transforms self-talk, and propels you toward your goals. They gave me the confidence to turn my doubts into determination, and my dreams into a reality I can touch.

If I could share one piece of advice with other heart-centered entrepreneurs, it would be this: embrace affirmations with an open heart. Craft your personal mantras, speak your truth, visualize your success, and watch as your mindset shifts. It's a journey that might feel a bit unconventional at first, but trust me, the magic lies in your unwavering belief. I've lived it, breathed it, and seen it shape my entrepreneurial path. It's time to let your doubts fade and your dreams flourish. Speak your success into being—one affirmation at a time.

KEY TAKEAWAYS

- Affirmations can be an important component of your success strategies.
- To be most effective, affirmations should be personal.
- Create affirmations to counter your doubts or limiting beliefs.
- Affirmations are not magic. They must be accompanied by hard work and purposeful action.

3

TRUST THE TIMING

ROBIN FITZSIMONS

I started my business, Wellness with Robin, as a brick-and-mortar location in 2011, leaving behind a full-time, secure but stressful job as a graphic designer. I had no clients to fund the overhead of my new space. Talk about fear! I had it in spades: insecurity, zero clients, no income, fear of being seen, and fear of failure.

Slowly, over the years since, my business grew. I loved creating and delivering in-person classes, where I could see my students in real life. I created lasting relationships with clients, who trusted me enough to refer their friends and family to me. I didn't invest in formal advertising and, honestly, I prefer referrals!

Some of my work like Reiki, I could deliver remotely. As my clientele grew, I was asked repeatedly to deliver more than Reiki sessions remotely. "You should deliver your classes online," was something I began to hear—and frequently. But I felt resistance. I mean RESISTANCE! I told myself that I was "technology challenged" and came up with all kinds of excuses for not growing beyond my local community. For about five years, I had a goal of taking my work online. For five years, I resisted. The fear of being seen by people I didn't really know took over the driver's seat—and I stayed on that ride for a long time. As an introvert, I wanted to stay where I felt comfortable.

Then I came to a place where I felt ready to grow. I don't know exactly what shifted, but I knew that, to really grow and make the impact I wanted to make, I had to put my (some of my) fears aside. One of my Angel Reading clients began a virtual assistant business (VA) working specifically with spiritually-led women entrepreneurs —like me. Having her guide and support me through the scary world of being online gave me confidence. She patiently taught me to do things that I'd been avoiding.

Over the years, I've been involved with women's networking and mastermind groups, which helped me push through some of my blocks. But it was my VA who opened me up by alleviating the fear of technology and taking that part on. To remove my blocks, I meditated and journaled using the energy of the full moon. I wrote down all my fears and then burned the paper in a fire.

That's when I had a huge *ah-ha* moment. If someone wasn't *interested* in what I teach or say, they won't watch me anyway! After all, that's exactly what unfolded in my brick-and-mortar business. I attract people who *want* what I have to offer. Once I realized this, growing my online business made sense; everything clicked into place! Truly, it all comes down to authenticity. When you're being authentic and consistent in what you're teaching, healing, or delivering, the right people will notice your messages.

I went back through my planners, journals, and mastermind notes, and everything I wanted was right there! I had newsletter topics, online class outlines, blog posts, and share ideas for social media. I got a new planner—I love a good planner, don't you?—just for my social media posts. I mapped out what I wanted to cover, like full and new moons, specials I wanted to offer, energy healing information, zodiac insights, and upcoming class offerings. I put a day aside to design my graphics in Canva, which I love to do (hello, I am a graphic designer!).

Planning

One nonnegotiable for me is a full day scheduled on my calendar just for planning. I learned this project management skill when I worked full-time as a designer. Having time to plan and design what I want to share is invaluable. It enables me to find additional content that I haven't created but is aligned with my messaging and suitable for my ideal clients.

When I'm planning the month's content—which includes my social media posts, newsletter, and blog posts—I think about what energy that month holds, I reflect on questions that have come from my students or clients, and I look at what classes I'm scheduled to teach.

Consistency

One thing that people often comment on is my consistency. I don't know about you, but when I follow someone's business on social media and they're inconsistent about what or when they're posting, I lose interest, and often unfollow them. One tool that helps me consistently post is Planoly, a social media posting tool.

Consistency is also crucial with newsletters. Before working with my incredible VA, I published newsletters inconsistently. And this behavior resulted in people unsubscribing. Simply by planning my newsletter content ahead of time and publishing it consistently, I've seen a significant increase to my email list, which means I'm getting in front of my ideal clients regularly.

Taking Classes Online

For me, this is where I had the biggest resistance. With my VA's encouragement, I located a mentor to support me in growing my online presence. I worked through the process of writing and recording my first class, Exploring Past Lives. Was it perfect? It was not. But neither are my in-person classes perfect. Do my students get what they need from my classes? Yes! They do. Consistently.

Here's the thing: I gave myself permission to be imperfect, to just get myself in motion. I know I can go back and rerecord if and when I choose to. If I'd waited until I had everything perfect, I still wouldn't have launched my online courses! I'm a Virgo, and may have a touch of perfectionism (ha ha!)—another thing I had to somewhat let go of.

Initially, part of my hang-up was that I didn't know how to do all of the things needed to take a class online. That was my fear showing up. Once I let go of needing to do everything myself, the right people presented themselves at the perfect time. And that, my friends, is called synchronicity!

It's Your Time

As you grow and market your business, you'll have fear rear its ugly head. That's normal. The things you fear are your greatest growth opportunities! Surround yourself with like-minded women's groups, join a mastermind, network (especially with other women entrepreneurs), listen to podcasts that align with your mission, and show up consistently. Leverage the Law of Attraction to bring the right groups and ideal clients to you—with ease and flow! Visualize having already met your hairy, audacious goals, despite any fears. Lean in, and know that *not* making a decision is, in fact, making a decision.

You've got this! I believe in you!

KEY TAKEAWAYS

- **Planning.** Schedule a full day to plan every month. Get a paper planner or an online calendar and map out what you want to share. Make this time nonnegotiable.
- **Consistency.** Look at how often you currently publish your newsletter and blog. Map out what you can

consistently deliver on—and keep to your schedule. Plan a writing day, if that's helpful. Jot down what people ask you about, as this will ensure you deliver what your ideal clients want to know. Easy peasy.

- **Go Online.** When you're ready to grow, decide to take your offerings online. Attract the right support and mentors, and do it scared! Making your offerings available in this way can increase your followers, email list, and impact!

4

LEARNING TO LEVERAGE FEMININE ENERGY

GAURI CHAWLA

As the child of first-generation immigrants from India to Canada, I learned quickly how to leverage my masculine energy. I attended a prestigious university in Canada, was recruited to one of the Big Four international consulting firms, and attended law school part-time in New York while working full-time. With my work ethic, it's no surprise that I made it to the C-suite in record time. My role was a global one, focused on developing and nurturing partnerships and alliances. Many times, I was the only woman at the table.

My company was sold in 2022, in part due to my team building successful, scalable partnerships that drove 45 percent of the company's revenue. I stayed through the transition, traveling to London, Sweden, and Norway regularly. I ought to have felt successful.

And part of me did. But mostly I felt exhausted, burned out. Honoring these feelings, I resigned. I realized that I had been successful because I worked using only my masculine qualities (something not uncommon with female executives in North America). In the male-dominated field of software sales, I unconsciously did what I needed to do to succeed. Masculine energy served me well up until a point. I created structure and delivered results.

I hustled, something I could only admit in hindsight. I doubted myself, hid, and mistrusted my instincts. As I spoke to other business-women, I realized I wasn't alone.

I also knew I wanted the rest of my life to be different. I wanted to work from a balance of masculine and feminine energy, to live a life that fills my cup rather than continuously draining it. Leaving my job gave me an opportunity to reflect on what worked for me and what didn't.

What I realized was, when I let my natural feminine energy flow, I inspired my teams, allowed creative ideas to develop, and nurtured others, which resulted in team buy in. But most of the time I was "doing" instead of "being." I was hard on myself and others. It didn't feel good.

When I balanced my energies, I had the best results—quantitatively and qualitatively. My team flourished. We helped our partners and alliances grow. We drove 45 percent of the company revenue through our partners. By enabling, supporting, and gaining the trust of our partners, they helped our customers succeed.

We created an incentive structure that created qualified sales leads from partners that grew opportunities which accounted for over 50 percent of the global qualified pipeline. Partner opportunities had larger deal values and a higher rate of closing. Our partners felt like an extension of our team to be able to provide candid feedback from our customers in order to help us improve our product gaps and to improve customer experience. I created a connected ecosystem based on collaboration, trust, relationships, and solutions.

My team was the best team in the company because they were focused, collaborative, informed, structured yet nurtured relationships, and were supportive of each other, the company, our partners and our customers. As a leader, I transformed from driving the team to inspiring the team. I became aware of how it felt to lead from my heart yet supported from my mind.

Masculine versus Feminine Energy Defined

As an individual, we have both masculine and feminine energies. Typically, in business, masculine energy dominates the landscape for both women and men. Regardless of vocation, one of these energies is usually the dominant for an individual.

Masculine energy is all about *doing*: externally focused, stable, logical, and predictable. Masculine energy traits are usually described as analytical, goal-oriented, driven, aggressive, structured, and confident. Masculine energy is about being in the mind.

Feminine energy is about *being*: nurturing, intuition, empathetic, receptive, creative, flexible, and vulnerable. Feminine energy is spontaneous and heart-oriented.

There is no doubt that masculine energy has its benefits. I can attest to that because I now realize how dominant my masculine energy was in business. It helped me get to where I am. This energy can also be aggressive, demanding, and—sometimes—destructive. Feminine energy can be inspiring, nurturing, and flowing. This type of energy can also be unpredictable and unstructured. Too much feminine energy can make one feel powerless or over-sensitive. There needs to be a balance between both energies within an individual, a team, and an organization in business.

New Paradigm: The Feminine

In the old paradigm, where masculine energy reigns supreme in the business world, an increasing number of entrepreneurs have recognized the power of feminine energy. Businesses with a strong foundation in feminine energy tend to be more sustainable and thrive over time. Feminine energy is supportive, nurturing, and creates an environment of collaboration.

In the old paradigm, we are conditioned to be productive and results oriented. Our entire world is focused on work + action = results. Creating systems focused on rewards for productivity is important, but it is not the only way to measure success in business. The masculine way of doing things can result in burnout, exhaustion, disinterest, and inevitable failure.

In a new paradigm, it's clear that feminine energy is just as powerful. In many cases, it's more powerful! Women often use their intuition and empathy to build successful businesses. What sets feminine-led businesses apart is their ability to create a safe and nurturing environment where employees, customers, and partners feel supported. This type of environment allows for creativity and risk-taking, which are essential for growth. Businesses with a strong foundation in feminine energy tend to be more sustainable and thrive over time. The easiest way to think of masculine versus feminine energy is to remember that masculine energy focuses on competition whereas feminine energy is collaborative.

By leading with feminine energy, I created a paradigm shift from a competitive approach where employees felt fearful and could only focus on short-term wins to a collaborative approach where everyone feels empowered and supported inside a community. On reflection, I see my feminine side as a leader. This is the part which I felt I had to hide and doubted because it was not the norm in the business. Yet, as I look back, this was the best stuff and the real reasons I got the unheard-of results. I move forward renewed, knowing I can trust my feminine side.

Marketing from feminine energy is different. It's no longer about cold-calling. It's about connecting with your customer or your ideal customer profile. It's about listening to what your customer needs and what their challenges are. Marketing from feminine energy is about tapping into those feminine qualities like empathy, authenticity, creativity and intuition.

It's about understanding your customer's pain points and ideal outcomes. When you truly empathize, your marketing becomes a bridge between your brand and your customers. Defining your value proposition and messaging is based on the value you bring to help your customers succeed. When you are rooted in that, your customers connect with you. With AI, there are even better tools to understand your customer and to create the content that attracts your customers to you. When that happens, then feminine energy creates the authentic connections. Authenticity builds trust, turning customers into delighted advocates.

Storytelling is a powerful feminine energy tool. Sharing stories that resonate with your customers inspires connection and action. Creativity is a hallmark of feminine energy that helps you think outside of the box and drive innovation that can translate into marketing campaigns, account-based marketing actions, and valuable events.

Feminine energy promotes collaboration across teams and organizations. When team members come together combining their strengths, innovation happens. Encourage teamwork and the creativity will flourish.

Feminine Energy Marketing is an invitation into a conversation and long-term relationship building through an authentic expression and respect rather than hard, "push them to sign" approach. Feminine energy inspires a purchase versus psychological tactics to make a sale. This way seeks resonance, compatibility, and authentic connection where a true win/win relationship can develop and last. And happy customers are the best marketing tool to speak to prospects and make them into customers and finally creating a community that loves your brand/product/company.

I do think it is important to mention having a balance between your feminine and masculine energies is vital to succeed and bring your best self to your role.

I write this in hopes to inspire you to trust your feminine side too. As you reflect on your feminine energy characteristics, which resonates most with you to allow to flow as a leader?

KEY TAKEAWAYS:

- Embrace and trust feminine energy in leadership.
- Openness to seeing feminine gifts to connect to quantitative results.
- Recognize feminine energy provides greatest growth for self, team, and organization.

5

ONE SIZE DOES NOT FIT ALL

GABRIELA BOCANETE

Anyone who has been in business for some time has heard what I consider masculine ideas, such as "work hard, play hard," "no pain, no gain," and "push out of your comfort zone."

Conceived by men, and replicated and promoted by the patriarchy, these ideas do not sit well with me. Same for working long hours, staying longer than your boss at the office, eating lunch at your desk, and pushing yourself to the limit for the infinite growth that greedy corporate models promote.

Specialist knowledge, skills, systems, best practices—these are all things that can be learned in school, university, or as an apprentice. In the case of entrepreneurs and freelancers, though, we learn by doing. We refine, repeat what has gone well, reflect upon and learn lessons, and grow with every experience.

Unlike the patriarchal ideas that permeate with corporations, having a business allows us to take a more feminine approach. How? By using ethics, self-knowledge, and perseverance. Let's take a look at each one to see how you can apply them to your business and marketing.

Let's Start with the Ethics

Someone once said ethics is universal; morals are parochial. They are similar, for sure, reminding us to be a good citizen, an honest human being, and doing the right thing in every situation with a clear conscience.

As a health coach, I do my best for my clients, updating my programs when new scientific information comes to light and leading by example in terms of healthy habits and lifestyle. While ethics as a corporate employee may equal productivity, as an entrepreneur, ethics means delivering what you have promised your clients.

Self-Knowledge

Know your talents, value your skills, and charge fairly for the contribution you bring to your clients. As a seasoned professional, your *presence*—the time you spend with your client—is incredibly valuable. Sometimes you don't even need to say much; your mere presence is a positive influence to the group.

Know your limits and respect your changing needs as your body matures. Learn how the body and the brain work, and don't expect peak performance if you ignore your body and brain's needs.

Self-knowledge also means having a sense of purpose and to adapt and accept that your purpose does not have to be forever the same.

Finally, have self-knowledge about an over-reliance on technology. We're all digital nomads some of the time, but that can lead us to working on holidays, weekends, and evenings. Thankfully this trend is shrinking, as mental health statistics show that social media and smartphone addictions cause sleep disturbances and other unwanted consequences.

Define Your Own Version of Success

Success is personal; you don't need to follow the business theories that lead so many people to burnout and anxiety. Write your own measures of success.

You're going to be successful even if you don't push yourself to a breaking point. As a woman, you instinctively know how to create and nurture lasting relationships with your clients. If clients then get the desired outcomes, these relationships can generate repeat clients and word-of-mouth recommendations.

Small is Beautiful

Consider a small, boutique-style business—perfect for the end of a career, transitioning to a healthier lifestyle, or meeting your family's needs. Or you may simply want to explore a new hobby and give yourself the chance to become creative in different ways. These are all valid reasons to keep your business small, just enough to stay engaged and interested in your field.

My boutique business is small and beautiful. I work with an intimate community of discerning clients, always updating my offerings to best serve them. I bring them new knowledge and they appreciate this. They love the small community feel, and they appreciate the chance to spend a few private minutes with me before I start the group sessions. I am there for my clients, and they appreciate the private conversation. I lift them up without breaking my back. It gives me joy, and it gives them a lot more value than what they paid for.

Recently, one client (let's call her Kim) shared how I helped her realize the importance of working less and prioritizing self-care. "I offloaded a few clients to younger colleagues," she said. Music to my ears! I knew that was the right decision for her. Here's why: Kim has a cognitively demanding profession as a conference interpreter. The advent of remote working for conference interpreters has been especially complex. The requirements to learn and use multiple new technologies while dealing with suboptimal sound quality has been challenging. Then, add in paying attention to multiple devices simultaneously while listening in one language, then understanding, processing, and speaking the same message in another language. Yes, Kim and her colleagues need to work less.

'Tis the Season to Slow Down

It's not only Kim. We all need to slow down. Go back to nature and respect her rhythms. Remember we are part of the natural world, just like our great-grandmothers were. They were seasonal and local, not only for grocery shopping and cooking, but also in the rhythms of life. They definitely spent more time outside, too.

We have lots more *stuff* these days, and we're being made to believe we still need more. This constant influx of stuff and information can be disconnecting us even more from nature as we increasingly lead our life indoors, in our built environment.

In a businesswoman's life there is seasonality, just like in nature. And once you've reached autumn and picked the fruit of your labor, you can slow down and give yourself space and time to reflect. What's the balance? What has it cost you in terms of health and well-being? Is it time to improve your fitness levels, to increase muscle mass for longevity?

Listen to your body, listen to your conscience; how much time and energy do you still want to spend working? Simplify your life and reduce your workload to make space for what's essential for you.

It may be that your children need you to help with childcare if you're a grandmother. You may want to spend more time with your retired husband who wants to take you traveling, to revisit favorite museums and galleries. To see those shows that you haven't had time because you've given priority to your work.

Of course, there are things you can delegate, using systems and a small team to keep your business going. Does it feel right? You will know in your heart. Do what is valuable for you, not what the business books say you should.

Small is Good for the Planet

One last thing: small ventures versus infinite growth models are kinder to our planet. Earth has suffered enough from excessive resource extraction and indiscriminate exploitation in so many areas. I've seen the destruction of habitats where orangutans used to roam in the isle of Borneo. The oil palm plantations have destroyed huge areas of native tropical forests and turned the apes into an endangered species.

Simplify and make your work an ethical and pleasant affair. That's what works for me. A small community of regular clients can support each other easier. Mine feels like a beautiful circle of belonging, nourishing for everyone involved. And it's international and multilingual, just like myself.

It gives me space to breathe, to forge collaborations and partnerships if I fancy. My work has been pioneering and an important influence in my professional associations. Now I get to look back, noticing how my contribution created new trends, making it possible for others to follow a double-track, portfolio career.

In fact, a younger colleague recently sent me a message to say she was following my example and invited me to attend her session at the conference where we're both speaking. I'm so fortunate to inspire people who are following in my footsteps, in what is not an easy path.

I think I deserve to slow down. And so do you, dear sister. One size does not fit all.

KEY TAKEAWAYS:

- As an entrepreneur, you don't have to follow the masculine energy ideals of working hard and hustling for your worth.
- A more feminine approach would be to consider your ethics, be aware of your self-knowledge, and write your own definition of success.
- It is okay to work less! In fact, your body and brain will thank you for it.
- How can you make your business smaller yet satisfying for you and your clients?
- You deserve to slow down.

FUNDAMENTALS

6

ATTRACT CLIENTS YOU LOVE AND LOVE THE CLIENTS YOU ATTRACT

LAURA TEMPLETON

Rebecca, a talented and driven coach, had been diligently serving clients for several years, building a successful coaching business. But deep down, she felt a sense of dissatisfaction, as if something vital was missing from her work.

Rebecca approached me, seeking my guidance as a brand communications specialist, to help her find clarity and alignment. While evaluating her coaching practice, it became evident that her branding and marketing strategies were not attracting the clients she genuinely enjoyed working with. Having marketed herself as a general business coach, Rebecca revealed that many of her clients didn't quite align with her values and goals, leaving her feeling unfulfilled.

Together, Rebecca and I embarked on a transformative journey to identify her true niche. We dug deep into what truly brought Rebecca joy in her work, and the values and goals she wanted her clients to share. It became clear that Rebecca's passion lay in supporting business owners who were dedicated to making a positive impact in their community, and who valued collaboration and teamwork.

As Rebecca wholeheartedly embraced her niche, a remarkable transformation took place. Her interactions with clients became more profound and meaningful. By attracting business owners who resonated with her values, Rebecca fostered authentic connections that went beyond the coaching sessions. She found herself genuinely loving her clients.

This newfound love for her clients had a profound impact on Rebecca's coaching practice. She approached each session with a renewed sense of purpose and enthusiasm. Rebecca's clients could sense her genuine care and dedication, creating a powerful dynamic that propelled them to achieve remarkable results. Through her love for her clients, Rebecca unlocked the potential within them, empowering them to reach new heights.

Word quickly spread about Rebecca's remarkable ability to connect with and love her clients. Her reputation as a coach who genuinely cared about her clients' success grew, attracting even more business owners who shared her values and goals. It was a testament to the power of finding one's niche and approaching coaching from a place of love.

As her business flourished, Rebecca found herself in a cycle of love and success. The more she loved her clients, the more fulfilled she felt, and the more impact she made. Her love for her clients fueled her drive to provide exceptional service and support, which, in turn, attracted even more clients who resonated with her values.

Through this transformative journey, Rebecca learned that finding her niche not only helped her align her business with her values and goals, but also enabled her to love her clients deeply. It was this love that propelled her to become a catalyst for their success, making a significant impact on their lives and the world.

Witnessing Rebecca's transformation, I realized the immense power of loving one's clients. It not only enhances the coaching experience but also creates a ripple effect of positive change. Rebecca's story serves as a reminder that when we find our niche and approach our work with love, we unlock a wellspring of joy, fulfillment, and profound connections with those we serve.

In any business, attracting clients is an essential part of growing and sustaining success. However, attracting the right clients—those that align with your values, approach, and business goals—can make all the difference in your work and overall satisfaction. Let's explore how to attract the clients you love and how to love the clients you attract.

Define Your Ideal Client

Before you can attract the right clients, you need to define who they are. Your ideal client is someone who is a good fit for your business, someone who values your work, is willing to pay for your services, and aligns with your mission and values. Take the time to reflect on who your ideal client is by considering the following questions:

- What are their demographics?
- What are their pain points and challenges?
- What are their emotional drivers?
- What are their goals and aspirations?
- What are their values?
- What do they expect from you as their service provider?

Once you have a clear idea of who your ideal client is, you can start to tailor your marketing and outreach efforts to attract them.

Create a Strong Brand

Your brand is what sets you apart from your competition and attracts your ideal clients. A strong brand is more than just a logo and color scheme; it's the essence of your business and what it stands for. Your brand should reflect your values, approach, and unique selling proposition (USP).

To create a strong brand, consider the following:

- Define your mission and values: What does your business stand for, and what values do you prioritize? Make sure your mission and values align with your ideal clients.
- Develop your USP: What makes your business unique? What value do you offer that your competition doesn't?
- Create a visual identity: this includes your logo, color scheme, and any other visual elements that represent your brand. Make sure your visual identity aligns with your mission, values, and USP.
- Craft your messaging: develop a clear and concise message that communicates your value proposition and resonates with your ideal clients. To learn more about crafting your message, consider purchasing my book, *30 Second Success: Ditch the pitch and start connecting!*

When you have a strong brand, you'll attract clients who align with your values and appreciate what you have to offer.

Build Relationships

Attracting the right clients isn't just about marketing and branding, it's also about building relationships. When you focus on building relationships, you create a loyal client base who is more likely to refer you to others and become repeat customers.

Here are some tips for building relationships with your clients:

- Communicate effectively: Make sure your clients know what to expect from your services and keep them informed throughout the process. Respond promptly to their questions and concerns.
- Be transparent: Be upfront about your fees and processes, and don't overpromise or underdeliver. Do the reverse,

under promise and overdeliver; this builds credibility, deepens connections, and often leads to referrals.

- Provide value: make sure your clients feel like they're getting their money's worth by providing high-quality services and going above and beyond their expectations.
- Listen actively: Take the time to listen to your clients' needs, concerns, and feedback. Use their feedback to improve your services and build stronger relationships.

When you build strong relationships with your clients, you create a positive reputation that attracts more of the right clients.

Love the Clients You Attract

Attracting the right clients is only half the battle—you also need to love the clients you attract. When you love your clients, you create a positive work environment that fosters creativity, productivity, and satisfaction. Here are some tips for loving the clients you attract:

- Set boundaries: Set clear boundaries with your clients to ensure you're not overworking or burning out. This can include setting working hours, communicating your availability, and limiting the scope of your services.
- Communicate effectively: Make sure your clients know what to expect from your services and keep them informed throughout the process. Respond promptly to their questions and concerns.
- Provide exceptional service: Make sure your clients feel like they're getting their money's worth by providing high-quality services and going above and beyond their expectations.
- Be empathetic: Put yourself in your clients' shoes and try to understand their needs and concerns. Show empathy and offer support when they need it.

- Show appreciation: Take the time to show your clients that you appreciate them. This can include sending thank-you notes, offering small gifts or discounts, or simply expressing gratitude for their business.
- Be engaging: engage with your clients on social media, by connecting, following, liking, commenting, and sharing when appropriate.
- Share your network: Invite them to join your network and offer referrals to the people in your network who can help them or who they can help. Simple introductions build stronger relationships.
- Learn from their feedback: use your clients' feedback to improve your services and create better experiences for them in the future.

Attracting the right clients and loving the clients you attract is essential to having a successful business. By defining your ideal client, creating a strong brand, building relationships, and marketing your services effectively, you can attract clients who align with your values and goals. And by setting boundaries, communicating effectively, showing appreciation, being engaging, sharing your network, and providing exceptional service, you can create a positive work environment that fosters creativity, productivity, and satisfaction.

KEY TAKEAWAYS

- Defining your ideal client is crucial for attracting clients who are a good fit for your business.
- A strong brand that reflects your mission, values, and unique selling proposition can help you stand out from your competition and attract the right clients.
- Building strong relationships with your clients through effective communication, transparency, providing value,

and active listening can lead to repeat business and positive referrals.

- Effective marketing through social media, content creation, attending networking events, offering incentives, showcasing your work, and collaborating with others can help you reach your target audience and attract the right clients.
- Loving the clients you attract by setting boundaries, communicating effectively, providing exceptional service, being empathetic, showing appreciation, being engaging, sharing your network, and learning from their feedback can create a positive work environment that fosters growth and success.

7

REFERRAL MARKETING: FROM CONNECTIONS TO GROWTH

STEPHANIE FRITTS

Let's venture into the world of referral marketing. The name itself is self-explanatory, but here's the definition according to the World Wide Web:

> Referral marketing is a marketing strategy that uses word of mouth and personal recommendations to grow a business's customer base. Unlike pure word of mouth, which happens organically, referral marketing programs are initiated and directed by marketers. The marketer's role is to influence the process and encourage existing customers to spread the word about a product or service by offering them incentives. Referral marketing is one of the least expensive marketing strategies that a business can use to grow its customer base.

Referral marketing is the power of your network becoming your business's most vocal advocates. So, who's your network? It includes:

- People you know (friends, family members, teachers, mentors, managers, past and present coworkers, etc.).

- People you meet (via networking groups).
- People who are already satisfied customers.

Why is referral marketing so powerful? Because when a customer is referred to you by a someone in their network, they'll come to you with a certain level of trust from the start. Heart-centered marketing, with its focus on authenticity, empathy, and building real connections, aligns beautifully with the ethos of referral marketing. In this convergence, referrals become more than transactions; they become stories shared with trust and care, setting the stage for a unique growth journey.

Crafting Your Heart-Centered Referral Strategy

The easiest place to start is with the humans you know (friends, family members, teachers, mentors, managers, past and present coworkers, etc.). I often tell the story of when I reached out (ahem, my virtual assistant reached out) to a list of 150 of my LinkedIn connections I know personally[2]—most of them were previous coworkers or executives I had supported, though a little voice in my head doubted they would be receptive after not hearing from me in more than fifteen years. I found that every person I connected with was excited to hear from me and learn what I was doing. And I truly wanted to know what they were up to as well!

Essentially, this is networking with people you already know, and the key to networking is giving as much, if not more, than you take in the interaction. Networking rarely yields immediate results (sales), but you're planting seeds that will grow and turn into sales. Case in point: one of the executives I supported twenty-plus years ago in the private equity industry reached out to me **six months after our initial call** to say that he had started his own firm and needed our services. Since then, he has referred two other clients!

LinkedIn is more than a professional networking site; it's a treasure trove for heart-centered referrals. The platform's potential lies in its ability to facilitate warm introductions. These introductions, backed by trust and authenticity, can be the bridge that connects your business with new opportunities.

Remember the phrase, "It's not what you know; it's who you know."? That's the essence of networking for referrals. Building genuine relationships through networking events and platforms creates a fertile ground for referrals to flourish. Engaging with others authentically fosters connections that can translate into referrals. That brings us to those you haven't met through intentional networking. As you read in the previous chapter on networking, you can network in many ways. I recommend joining groups that resonate with you, where you can be your authentic self, and where your ideal audience hangs out. Some examples of groups I belong to are Ellivate Alliance, a women's entrepreneurs' community of brave women entrepreneurs who are channeling our feminine wisdom to run our businesses, create social change, and find joy in who we are and what we do; a free virtual networking tool called LunchClub; and last but not least, the wonderful virtual women's networking community, Virtual Networkers. I've also networked on webinars when appropriate.

Lastly, we have humans who are already your clients. Have you ever had a customer who raves about your business as if it's their favorite ice cream flavor? That's the kind of loyalty we're aiming for. In a referral, trust is the currency that matters most. Think about it: When a friend suggests a product or service, you're more likely to try it because you trust your friend's judgment. Building and nurturing this trust is where your heart-centered approach comes into play. It's about showing up authentically, delivering on promises, and being a business that can be relied upon. Creating exceptional customer experiences is like laying the foundation for these heart-centered referrals. It's about understanding your customers, exceeding their expectations, and planting the seeds for those genuine recommendations.

But what about referral programs and/or affiliate programs?

These are great, too, as long as you have the capacity and tools to manage the program itself. Some software tools can manage these programs, but you'll need the budget for the tool and someone to set it up for you. You could DIY, but I don't recommend it unless it is in your zone of genius! I'll just say this: most, if not all, of your contacts will be more than willing to send you referrals, no strings attached.

Emotional Connections: The Referral Glue

Emotions have an uncanny way of sticking with us. This is why some television commercials bring us to tears! A heart-centered referral does not just provide a solution; it tugs at heartstrings. Stories that evoke emotions create a connection that's hard to forget. When your referrals are backed by genuine stories that resonate, they become more than just recommendations—they become personal endorsements that touch the hearts of potential customers. And speaking of personal endorsements—always ask your happy clients for testimonials! If they're willing to refer you, they will publicly endorse you. These testimonials can be used on your website, social media, newsletter, etc. Just make sure you get permission from the client first.

Turning Heart-Centered Referrals into Action: Igniting the Referral Spark

So, you've got happy customers who love your business—it's time to fan the flames of referral excitement. This isn't about pushing people to refer; it's about creating an environment where they can't wait to share their positive experiences. It's like creating a referral wildfire (in a good way). Some clients will send referrals naturally; others may need a reminder or a quick and easy way to refer you. Others will need a nudge because they're busy humans! In that case, you'll need to *ask* them for referrals, and you'll want to make it as quick and easy as possible for them. If you use a CRM (client/customer relationship manager), you can set up a lead capture form that your clients can send to the referral. You can also encourage your clients to simply make an e-troduction (email introduction) to that referral—one tip I recently learned was to create your own e-troduction blurb for them to make it even easier.

Sustaining the Referral Flame

Gratitude is a heart-centered superpower. Your referrers are your business's unsung heroes, spreading the word because they genuinely believe in what you offer. Celebrating their efforts with heartfelt thank-yous, personalized gestures, and a virtual high five goes a long way in showing appreciation. Ask them if they would be okay with you thanking them publicly on social media.

You've got referrals, but the journey doesn't end there. Follow-ups are like keeping the campfire alive after the initial spark. Staying in touch, listening to feedback, and continuing to provide value to both referrers and potential customers strengthen the relationships you've built.

Tackling Referral Challenges with Heart

Referral marketing has challenges, but a heart-centered approach is your shield. Embrace challenges as learning opportunities. Maintain your commitment to authenticity and empathy even in the face of hurdles. This heart-centered resilience helps you overcome obstacles and deepens your connections.

Growing your business is exciting but staying true to your values is paramount. A heart-centered business embraces growth while ensuring its core remains unwavering. Balancing expansion with maintaining the authenticity of your referrals is where your heart-centered compass comes into play.

Afterward

When infused with heart-centered authenticity, referral marketing isn't just a strategy—it's a journey of building connections, one referral at a time. By blending trust, empathy, and a dash of care into your referral mix, you're not just gaining leads but creating a community of loyal supporters. So, grab these heart-centered insights, sprinkle them into your referral strategy, and watch your business flourish. Because referrals? They're your not-so-secret sauce for growth, wrapped in the warmth of genuine connections that make your business journey extraordinary. Your business adventure just got much more exciting—it's time to embark on the heart-centered referral voyage!

KEY TAKEAWAYS

- Referral marketing is super powerful because your leads are coming to you with more trust than if they found you through traditional marketing.
- Your network of friends, family, and former coworkers is a great place to start because they want you to succeed!
- Networking is all about building relationships and finding synergies, not getting an instant lead or sale.
- Most of your satisfied clients will shout from the rooftops about your service or product; if they don't, they may just need a nudge!
- Be your authentic self, lead with your heart, and extend gratitude to those who send referrals—The Universe will respond accordingly.

8

HOW TO HAVE COFFEE DATES THAT DON'T WASTE YOUR TIME

JILL CELESTE, MA

I watched Bobbie jump out of her car, a laptop cradled in her arms. She rushed into Panera Bread, where I was waiting for our coffee date. Bobbie was late, and my "spidey senses" were on high alert.

I didn't want to go on this coffee date—or any coffee date. They seemed like a grand waste of my time, and Bobbie's tardiness added to my trepidation.

Bobbie waved and smiled at me as she entered the lobby, making a beeline to the cashier so she could place her order. Within minutes, she sat across from me, beads of sweat on her brow.

"I am so sorry that I am late," she exclaimed, opening her laptop.

"No worries. Let's get started with our coffee date," I replied, forcing a smile.

Bobbie typed something into her computer, and then flipped it around so I could see the screen. It was her Facebook page.

"I am so glad we could meet in person. I have some Facebook marketing questions to ask you," Bobbie said. Not waiting for a reply, Bobbie launched into her questions.

Forty-five minutes later, I stood up and told Bobbie that I had to pick up my kids from school.

"Jill, thank you so much for the coffee date!" Bobbie said with a smile.

I gritted my teeth. *This wasn't a coffee date.* This was a free coaching session where I answered Bobbie's Facebook questions. She didn't stop talking long enough for us to discover how we can help each other.

When I got back to my car, I wanted to cry. My suspicions had been correct. I spent more than an hour of my time with someone on a "coffee date," and it felt like a waste of time. I could have been doing client work and earning money!

I threw my car in reverse, vowing never to do a coffee date again.

Changing My Coffee Date Mindset

Intellectually, I knew that banning coffee dates was not the answer, but I never wanted to have a "coffee date" like the one I had experienced with Bobbie.

Thankfully, shortly after this coffee date, my business mentor taught me how to have effective coffee dates—not only the "how" but the importance of a positive coffee date mindset.

I was reluctant about having coffee dates, worried that I was wasting my time. Well, that became a self-fulfilling prophecy. If I am reluctant and worried, I wouldn't have a positive result from *any* coffee date.

With my newfound knowledge and mindset, I started coffee dates again. Guess what? Every coffee date was amazing! Did each one result in referrals or clients? No. But each one did result in new relationships, expanded visibility, and overall good vibes.

Fast forward to now: I am the founder of Virtual Networkers—a global virtual networking organization for women entrepreneurs—and I always encourage my members to have coffee dates, especially with each other. I even hosted a Summertime Coffee Date Challenge, where I awarded prizes to members who held coffee dates over a three-month period.

My Virtual Networkers met the challenge, too! They collectively held *hundreds* of coffee dates. They reported so many benefits from their coffee dates, from business collaborations and referrals, to deepening relationships.

And that's what I want for you, too.

Are Coffee Dates for You?

You may be thinking: *Wow, this sounds so old-fashioned? Holding thirty-minute conversations over coffee?*

Yes, coffee dates may be old-fashioned, but they stand the test of time *because they work*, especially with the right knowledge and mindset.

Perhaps you have objections, like I once did:

- You are not sure how to make coffee dates worth your time.
- You are not sure how to ask another business owner to join you for a coffee date.
- You are an introvert and need to conserve your energy.
- You don't mind coffee dates, but you can't spare more than thirty minutes, and you are afraid it will go longer.

I am living proof that you can overcome any objections to coffee dates—and actually learn to like them! As I mentioned, you just need the right knowledge and mindset.

Let's Start with Mindset

What helped my mindset around coffee dates was knowing I had a firm process about how the coffee date should go. Perhaps it was a false sense of control, but I felt better knowing I had an agenda I could use for coffee dates.

Hand in hand, I approached coffee dates from this perspective:

how can I build a relationship with this person? Notice I didn't say *how can I get a referral,* or *how can I get this person to buy from me?* While coffee dates can lead to sales and referrals, the first goal of a coffee date is about building relationships. Embracing this perspective takes the pressure off, and honestly, leads to better results.

Now, I will share the process I used for coffee dates—but with one caveat. You will notice that the conversation centers on sales and referrals, and that's okay! You can talk business while building relationships. And feel free to sprinkle in the personal stuff, if it feels right to you.

Asking for a Coffee Date

If you identify someone you would like to meet for a coffee date, here's what you should say: *"I believe I can refer clients you. Would you like to meet for thirty minutes, and discuss our ideal clients and see if we can help each other?"*

The other entrepreneur will say "yes," and you take it from there. The beauty of this request is that it's about the other person. Who wouldn't want more clients, right?

Identify How You'll Meet

As an introvert, I only have virtual coffee dates. They are easy and effective, and they save time, money, and energy. With that said, you may prefer in-person coffee dates—where you meet at a local coffee shop or restaurant—and that works well too.

If you choose to go the virtual route, you can leverage video conferencing tools such as Zoom, Facetime, or Skype, so you can see the person while you're meeting virtually. If none of those options are available, you can always call each other.

What to Do During the Coffee Date

Before you start your coffee date, make sure you have information about your ideal client. I would recommend creating a one-page flyer that outlines who your ideal client is. This is an easy thing to email your referral partner once the coffee date is over.

Next, as soon as you start your coffee date, remind her you are eager to learn more about her ideal clients—and you want to share information about your business too. Here's how I usually state it:

"I am so excited to learn more about your business today! I don't want to keep you too long. Why don't we talk about your business for about fifteen minutes and then switch over to mine?"

This sets the stage that this is a *mutual relationship* and keeps the conversation on pace. Don't be shy about setting a timer either.

While your coffee date talks, make sure you get a good idea of who is her ideal client and what type of business she is looking for. Take notes, if needed. If you know of someone who is her ideal client, share that person's name and promise to make an introduction (do it then, if you can, through email or text message).

Then, when it's your turn, *specifically* ask your coffee date if she knows anyone who is like your ideal client. If she does, ask her to make an introduction, and be sure to write down the names of anyone she thinks is a good fit.

After the Coffee Date

The post-coffee date is super important, and it's often a step that's forgotten or skipped by entrepreneurs. Once the coffee date concludes, immediately email your new referral partner:

- Thank her for her time.
- Summarize who her ideal client is and the names of the people you think may be a good fit. Mention that you'll reach out to these people right away.

- Gently remind her to reach out to anyone she knows that is your ideal client (if she provided names, list them). Send your Ideal Client flyer as an attachment.

After sending this email, make those introductions you promised. *Do this right away.*

Now, go to your calendar, and think about *when* you want to reach out to your new referral partner again. For example, plan to do another coffee date quarterly and maybe a reach-out email monthly.

Your job, at this point, is to keep your name top of mind. People have great intentions, but unless you wave your hands around and make them remember you, you may not get any referrals. Be diligent in following up with her referral suggestions. And be sure to keep in touch with her—well after the coffee date.

Now that you have the process, invite someone today to do a coffee date with you. If you can have at least one coffee date a week, you are making friends, increasing your visibility, and putting more "ears to the ground" for you. With time and patience, you will see that your coffee dates help your business (and life) so much!

KEY TAKEAWAYS

- Coffee dates can help you grow your business in a heartfelt way.
- Having a coffee date process will prevent any "wasted time" fears you may be concerned about.
- Embracing the right mindset about coffee dates is critical. If you go into coffee dates expecting a negative experience, you will probably have a negative experience.
- Focus on building relationships. The sales and referrals will come, but first you must build a relationship with your coffee date.

- Don't forget to follow up with your new referral partner after your coffee date. Keep your name top of her mind.

9

HOW TO DIFFERENTIATE YOURSELF IN A CROWDED MARKETPLACE

CLARE WHALLEY

Female business owners can often be the last people to have confidence in their ability. Somehow not being able to see, hear, or feel the same capabilities their clients *know* and love them for. This is where our male counterparts differ. This well-known yet anonymous quote sums it up perfectly: "carry yourself with the confidence of a mediocre man."

Without the recognition of your unique skills and strengths, it's impossible to project the level of confidence *needed* to differentiate yourself in any marketplace, let alone a crowded one. *Why should someone choose to work with you over the next person delivering the same or similar service? What do you and your business offer that's different to your counterparts?*

If you're struggling to answer these questions, over the next few pages, I'd love to turn this conundrum on its head for you. Let's help you truly differentiate and stand out from the crowd.

And let's be clear—exuding confidence is not about arrogantly shouting from the rooftops how brilliant you are and how people are queuing to work with you. Self-confidence is about being able to share, with assurance, whom you work with and how, and what results you help others achieve.

Are you ready to become a more self-assured you so you can confidently differentiate you and your business?

What's Your Superpower?

Being a business owner means you will already be well acquainted with stepping outside of your comfort zone—from attending your first networking event to posting on social media. For first timers this can be scary stuff. And you will (hopefully!) have learned nothing disastrous happened as a result of doing something new. We know good things happen when we develop.

The same rules apply when we differentiate ourselves and our business. It can (and does!) feel intimidating when we step outside of the norm, but the very definition of "differentiate" showcases how right it is for us to grow and *be* different:

> **Differentiate (verb):** make or become different in the process of growth or development.

So how do you establish your superpower? That thing you are gifted at that you can use to demonstrate your differences from the next person and be easily recognizable for what you do?

The "What's your Superpower" exercise is taken from *The Big Leap* by Gay Hendricks. Hendricks shares the concept of working within four different zones: the zones of incompetence, competence, excellence, and genius. The very nature of running a small business leads to the wearing of many hats, such as marketing, accounts, customer service, client delivery, and advertising to name a few. And most business owners wear many, if not all the hats!

Now, intellectually we know it's impossible to be excellent or genius at them all, so as a result, business owners are involved in tasks that sit firmly within their zones of incompetence and competence a higher proportion (probably around 50 percent) of the time. The result? Mediocrity. Ouch.

Working out what you *love* to do and what tasks you do extremely well will help you get clarity on where your time is currently being spent versus where you want your time to *be* spent, i.e., helping you move toward working more in your zones of excellence and genius.

This exercise will help you look objectively at all the tasks you complete to see which of them could be better placed elsewhere. Perhaps empowering a team member or finding someone else, such as a virtual assistant (VA) whose skills lie within that area of expertise.

Start by noting all the activities you have undertaken in your business (in the last month) and categorize them using this key:

- **Zone of Incompetence.** This is made up of all the activities you're not good at.
- **Zone of Competence.** These are the things you're competent at, but others can do them just as well.
- **Zone of Excellence.** These are the tasks you do extremely well.
- **Zone of Genius!** This is the set of activities you are uniquely suited to.

The best way to approach this exercise is to get a trusted friend and colleague (someone else also in business) to do this with you. Completing this exercise with another entrepreneur helps us see what we are truly great at.

The key to success is to be working within your zones of excellence and genius at least 80 percent of the time. You can only achieve this by ditching the tasks and activities you're not so competent at. These tasks are taking up too much time and winning your attention away from what you are truly meant to be doing—exercising your excellence and genius. Herein lies your *differentiation*.

Penny, a business owner of ten years, worked through this exercise and found even this far into her business, she was working within her zones of incompetence and competence 65 percent of the time, only 30 percent of it within her excellence and 5 percent in her genius.

The impact? Well, tenfold. Penny was busy, hassled, and discontented. She felt like she was never doing or being enough in her business (let alone her family life). The unfocused approach to running her business impacted her marketing message too; it was inconsistent and mixed. Penny's mind was frazzled, her time squeezed. She had no time or headspace to grow and little to no confidence to increase her pricing structure. No surprise when 65 percent of her time was spent in mediocrity. Penny felt stuck.

Completing the exercise helped Penny see where immediate changes could be made to start turning her mindset and business around.

To Niche or not to Niche?

By talking about niching, we can just as easily be saying differentiating. Business owners often get very nervous when discussing the potential of niching, sharing fears such as: I won't have enough business, or I don't want to turn business away or my existing clients (outside of the niche) will no longer want to work with me.

This way of thinking simply isn't true. All your favorite brands are probably niched. They've just done it in such a way you feel they are for everyone. Take Starbucks and their mission statement: "To inspire and nurture the human spirit—one person, one cup and one neighborhood at a time." Now we know Starbucks meets many coffee needs; however, they've made themselves feel they are all about *you*.

Check out Jane's story, how she sees herself having a *broad* niche and how she's managed to differentiate (and niche) herself successfully.

Before starting her own business, Jane worked at an advertising agency and was always the one to take charge. She had a natural confidence in speaking her mind, but it was always on a practical level—problem and solution. All the makings of a business owner!

Jane's business started well, and she picked up work from a few agencies who knew her, but for Jane, working directly with clients was what she wanted to do. This was pre-websites and LinkedIn, so word of mouth was her greatest weapon.

First, Jane made the decision to never refer to herself as freelance. She was a mini- agency able to offer a one-to-one service. She developed a broad niche; working with KBB companies (kitchen, bedroom, and bathroom) who sell via distribution through third parties.

Now thirty years into her marketing business, Jordan Twigg, Jane shares of her decision to niche: "It gives me a broad spectrum of work and it means I get to know what works and what doesn't. I also walk away from potential clients that don't fit with the way I worked or had products I couldn't believe in."

Niching doesn't have to be a massive U-turn on who you are and what you deliver. And it doesn't have to feel like you're deciding that you can't go back on. Like Jane's, it is a decision made from confidence—in you and your ability. Marketing your business is trial and error. See what works for you.

Read how Jenny Procter of Bondfield Marketing got comfortable with niching her business: "I attended a women's networking group and each month a member gave a presentation about her business. One day I went along and listened to a presentation from another marketing consultant. She was completely unlike me. She was energetic and lively. She was also a bit sweary and sassy. I was despondent about what this meant for my business—why would anyone want to work with me, when they could work with her?"

Jenny shared her fears with a supportive business coach also in the room who had a completely different take on the situation. The coach shared that while Jenny had been thinking the person was better than her, the business coach had been looking around and while many people were clearly attracted to her style, she felt at least half of the ladies would not choose to work with her because she just wasn't the marketing consultant for them.

This insight enabled Jenny to take a leap of faith and contact the energetic marketer. They had coffee and compared businesses with Jenny quickly realizing that they were very different people, whose skills were suited to very different businesses. She was an extrovert, whereas Jenny's ideal customers were introverts. "And from that original moment of comparisonitis, 'Marketing for Introverts' was born!"

So How Do You Create Your Niche?

Know your niche will not always come to you like a light-bulb moment. Oftentimes it takes thought, planning, and time.

Think about all the people you have *loved* working with over the last eighteen months to two years. What are their commonalities? Find out why they chose to work with you. What did they love about working with you?

All these answers will help you to build up a clearer picture of what you *uniquely* offer. Go on, be bold. Differentiate yourself; niche your business. Your marketing content will love you for it!

Lots of Love,
Your straight-talking, no-nonsense business coach for creatives
(see what I did!?)

KEY TAKEAWAYS

- Taking the time to recognize your unique strengths and skills, "Your Superpower" will help to empower you, grow your confidence, and will make it easier to differentiate yourself in a crowded marketplace.

- Niching can be simpler than you think. Niching can be about your or your clients' commonalities rather than a particular product or service you offer.
- Being bold, stepping outside of your comfort zone and sharing your inner thoughts and fears, with the right people, develops clarity and can create business-affirming outcomes.

10

CRAFTING A MISSION-DRIVEN MARKETING JOURNEY

STEFANIE JOY MUSCAT

In the dynamic landscape of business, a well-defined mission and steadfast values serve as the compass guiding not only nonprofit organizations but also entrepreneurs navigating uncharted territories. These elements illuminate the path to sustainable success by fostering authenticity in a fiercely competitive market. This chapter embarks on a transformative voyage to unearth the intricate process of identifying an entrepreneur's mission and values, while further exploring the art of seamlessly weaving these fundamental elements into innovative marketing strategies.

Setting Sail: Navigating Mission and Values

Embarking on Self-Discovery. The entrepreneurship voyage commences with an introspective journey that unravels the tapestry of an entrepreneur's motivations, passions, and ambitions. This introspection forms the bedrock upon which the mission of the venture is etched. Take, for instance, the case of Patagonia, an outdoor clothing company. Yvon Chouinard, the founder, combined his love for nature with a deep commitment to environmental sustainability, leading to a mission of not only creating high-quality outdoor gear but also inspiring and implementing solutions to environmental crises.

Forging Core Values. The lodestar of any successful enterprise lies in its core values—the guiding principles that underpin its culture, actions, and decisions. These values encapsulate the entrepreneur's essence and principles. Apple Inc., for example, embodies innovation as a core value. Steve Jobs, Apple's co-founder, forged a legacy of disruptive innovation by challenging the status quo. The company's products, branding, and marketing strategies are imbued with this value, allowing customers to associate Apple with groundbreaking technology and design.

Reading the Market Winds. The journey towards entrepreneurial triumph calls for more than just personal conviction; it entails an astute understanding of the market's currents. Entrepreneurs must not only be attuned to their inner callings but also identify market gaps or challenges. TOMS, the shoe company, exemplifies this principle. Blake Mycoskie founded TOMS with a mission to provide a pair of shoes to a child in need for every pair purchased. This mission not only addresses a social concern but also resonates with consumers seeking to make a positive impact through their purchases.

Crafting the Narrative: Weaving Mission and Values into Marketing

1. The Tapestry of Authenticity:

- *Threads of Consistency*: A mission and values serve as the threads weaving a tapestry of authenticity. This consistent narrative fosters trust and cultivates loyalty among customers. Consider the success of Airbnb; their mission of "belonging anywhere" is meticulously woven into every aspect of the company's interactions and offerings, resonating with travelers who seek unique, local experiences.

- *Narrative Spinning*: The power of storytelling becomes evident when entrepreneurs spin their personal journey into a narrative that humanizes the brand. The "Ben & Jerry's" ice cream company, founded by Ben Cohen and Jerry Greenfield, embarked on a journey of social responsibility. Their mission to make "the best possible ice cream in the nicest possible way" is not just a statement but a narrative that consumers connect with, as they see the company's commitment to sustainable sourcing and social causes.

2. A Connection Forged in Values:

- *Anchoring in Shared Values*: An entrepreneur's marketing strategy can establish a deep connection by anchoring in shared values. Tesla, led by Elon Musk, exemplifies this. Their mission to accelerate the world's transition to sustainable energy resonates with environmentally conscious consumers who share the value of combating climate change. Tesla's marketing isn't just about selling electric vehicles; it's about catalyzing a global movement.
- *Segmentation's Mosaic*: Through market segmentation based on shared values, entrepreneurs can craft tailored marketing messages. For instance, the outdoor gear company REI appeals to both outdoor enthusiasts and environmental advocates. By focusing on the shared values of adventure and sustainability, REI creates marketing campaigns that resonate with distinct segments of its audience.

3. Narrative Nourishment: Crafting Content with Care:

- *Knowledge Dispersion*: Sharing insights and knowledge related to the mission and values can position the brand as an authority. A prime example is Whole Foods Market, which thrives on promoting health and organic living. Their marketing content includes educational materials about nutrition, sustainability, and sourcing, establishing them as a trusted source of information in their industry.
- *Cause-Driven Creation*: Brands whose mission aligns with a social or environmental cause can create content that showcases their commitment. Warby Parker, an eyewear company, not only provides affordable eyeglasses but also focuses on addressing the global issue of visual impairment. Their marketing campaigns often highlight the impact of their Buy One Give One program, which provides glasses to those in need.

4. Harvesting Collective Wisdom: Partnerships and Collaborations:

- *Harmonizing Values*: Collaborations with like-minded entities amplify the brand's message. An example is the partnership between Starbucks and (RED), an organization fighting against HIV/AIDS. This collaboration seamlessly aligns Starbucks' core values of social responsibility and community engagement with (RED)'s mission, creating a win-win situation.
- *Bridging the Divide*: Collaborations can lead to innovative projects that reinforce the brand's commitment to its mission. The partnership between Adidas and Parley for the Oceans is a stellar example. Their shared value of environmental conservation led to the creation of sneakers made from recycled ocean

plastic, a testament to their joint commitment to sustainability.

5. Sailing the Ethical Waters:

- *Transparency, A Beacon of Trust*: Ethical marketing thrives on transparency. The Body Shop, known for its commitment to cruelty-free and ethically sourced products, maintains an unwavering commitment to transparency. By openly sharing their sourcing practices and philanthropic initiatives, they establish trust with their customers.
- *Beyond Lip Service*: Entrepreneurs upholding values like sustainability must ensure their marketing reflects genuine efforts. Patagonia goes beyond lip service by not only creating high-quality, long-lasting products but also encouraging customers to repair and reuse items. Their "Worn Wear" initiative not only aligns with their sustainability mission but also empowers customers to adopt a more conscientious consumption pattern.

Navigating the Challenges: Anchoring Mission and Values

1. Safeguarding Authenticity:

- *Challenge of Drift*: As businesses grow, they might unintentionally drift from their original mission and values. Starbucks, as it expanded globally, faced this challenge when maintaining the local community connection became harder. However, the company introduced initiatives like the "Starbucks Reserve Roasteries," which bring back the sense of community and craftsmanship that resonates with their original values.

- *Constant Course Corrections*: To ensure that the mission remains the guiding force, regular reevaluation is essential. Companies like Google, while adapting to evolving market trends, consistently reiterate their "Don't Be Evil" motto, emphasizing their commitment to ethical practices.

2. Adapting to Changing Currents:

- *Navigating Change*: The business landscape is dynamic, demanding agility. Coca-Cola, originally centered around a beverage-focused mission, adapted to health-conscious consumer trends by diversifying their offerings to include healthier beverage options, demonstrating an alignment with evolving societal values.
- *Balancing Act*: While adapting, the core values must remain unwavering. McDonald's, for example, introduced healthier menu options while still maintaining their commitment to affordability and convenience, aligning with their value of serving families from diverse backgrounds.

3. Finding Equilibrium Between Profit and Purpose

- *Challenges of Trade-offs*: Balancing profit with a genuine commitment to the mission can be daunting. Yet, successful companies like Ben & Jerry's prioritize both. Their social mission did not deter them from growing into a successful brand; rather, it became an integral part of their success story.
- *Holistic Perspective*: Viewing profit as a means to sustain the mission, entrepreneurs can communicate how their business growth contributes to the greater good. For instance, Toms Shoes not only donates shoes but also

invests in other sustainable solutions through its One for One model.

What actions can you take today, to undertake this journey with your business?

Entrepreneurs can begin by crafting a resonant narrative that intertwines their personal journey with the brand's mission. Sharing how their mission stems from personal experiences creates an authentic emotional connection. Integrating core values into marketing messages strengthens the brand's identity, while social media platforms serve as avenues to share mission-related content and engage with the audience.

Further, creating educational content, such as blog posts or videos, educates the audience about the issues the mission addresses, establishing the brand as an authority. Collaborations with like-minded organizations and influencers expand reach and reinforce the mission's impact. Customer testimonials offer social proof and resonate with those who share similar values.

Finally, transparency is key; showcasing sustainable practices and behind-the-scenes efforts builds trust. Email campaigns and cause-related initiatives directly involve the audience in the mission's progress. Encouraging feedback and engagement shows that customer input shapes the brand's path, forging a lasting bond with those who champion the same values.

An entrepreneur's marketing prowess blooms when anchored in a compelling mission and fortified by steadfast values. By adeptly intertwining authentic storytelling, community-building, and an unflinching dedication to shared values, entrepreneurs forge a connection that reverberates with customers, driving enduring success. As consumers increasingly seek authentic, purpose-driven brands, aligning mission and values with marketing strategies becomes not only a choice but a survival imperative. Just as seasoned mariners navigate by the constellations, entrepreneurs chart their course through the radiant guideposts of their mission and values, propelling their ventures toward purposeful achievement.

KEY TAKEAWAYS

- Unearthing Mission and Values: Entrepreneurs commence by introspecting their motivations, leading to the venture's mission. Core values become guiding principles shaping a company's culture and decisions. Understanding market trends is crucial for entrepreneurial success.
- Authentic Narrative Weaving: A brand's mission and values form the threads of authenticity, fostering trust and loyalty. Connecting with consumers through shared values creates a global movement. Sharing insights positions a brand as an authority.
- Navigating Challenges with Purpose: Safeguarding authenticity as businesses grow is vital. Adapting to changing consumer trends while upholding core values demonstrates agility. Balancing profit with mission commitment shows that a social mission can be integral to a brand's success.

11

MAKE THOSE CONNECTIONS THROUGH NETWORKING AND REFERRALS

JOANNE FLYNN BLACK

In this book, you will read of the many ways to attract new clients. And you should try some! In addition to those things, you should always use networking and referrals as an excellent way of bringing in new clients. I have! And it's gotten me where I am today. Let's explore how it's done.

Networking

Networking sometimes gets a bad reputation because it sometimes is, well, bad. You may wind up with everyone interested in pitching their own stuff that they don't even listen to what you have to offer. However, it doesn't have to be that way. I was invited by one of my friends to be a guest speaker for a Virtual Networkers Community even though I was not part of the group. My friend said great things about the women there.

So, I joined the meeting one week to give my presentation on "The Power of Community." I was blown away within the first five minutes during introductions, because everyone had their thirty-second introduction of what they did and who they helped. They were so clear. I was especially impressed because I had just gotten off a call where the person I was talking to, took fifteen minutes to explain what he did. We barely had time to talk about business.

But what really impressed me about this group was, at the end of the session, each person had a minute to promote something they were offering. And in addition to the promotions, there were "shout-outs" to others in the community. And lots of referrals. "Hey Mary, I came across someone who was looking for exactly what you do, and I passed on your information."

"I have a lead for you," or "I know someone who would be a perfect client for you," were common sentences I heard. I was so impressed how people cared not only about promoting themselves, but also about promoting each other. So, I became a member. And from that group alone, I gained several clients. These were people who needed exactly what I was offering. And our interactions never felt "salesy." Each connection felt like she found me at the time she needed me. It felt aligned.

Also, I'm a client of some members there too. Now when I think of something I need for my business, I always check the Virtual Networkers community first. "Is there anyone here that does XYZ?" is my first question. Usually there is. If not, they refer someone else they know. And I trust their opinion. Because they know me and know who I would work well with.

Referrals

I'm a big believer in referrals. Because if someone I know recommends someone else, I'm more apt to trust that person. She comes recommended. And I know my friend wouldn't refer someone she didn't love.

The easiest calls I have are when I get referred by someone. I don't need to spend time proving myself and my capabilities. The conversation usually goes something like this, "Debby referred you and I think her community is great. How can you strategize and build one for me?"

And then we launch into more detail about the client, as opposed to me talking about myself.

Referrals can come with nothing expected in return or you can offer to someone that if they have referred you and you get a paying client, you can give them something. I'd be sure that it is genuine and that they really are referring you because they believe in you and your work, not only because they want something from the sale.

This "something" can be in the form of cash or services. As an example, you could offer a free hour session with you to anyone who refers you who then becomes a client. Or you can offer them something financial. For example, if someone refers a $5,000 client to you, you can offer them $500.

They're happy to get the money and you're happy that you have a new client. Or you can offer them an option. You can say, "Hey, I'm offering an hour of my service or $500. You decide." It is important to remember that you only want to do this with someone that you feel aligned with and connected to. You know they will bring good people to you.

Plus, you don't want someone referring you just to get the bonus. Rather, make the bonus exactly what it sounds like: something special for taking the time to think of you and spread the word.

Now when I'm on a group call and I mention what I do people will remember. I had a woman recommend me after we were in a group session together. Other than meeting her online a few times, we didn't really know each other. But she knew what I did, she liked my energy, and knew that I would be a good resource for her friend. She recommended me, her friend and I got together, it was an instant match, and she became a client right away.

Sometimes I even get referred by the software company that I use to implement and grow communities. Since they don't do what I do, they send the customers who want assistance to me. And those are the easiest of easiest calls. "Hey, if Mighty Networks is referring you,

you must be good!" is something I hear often. This didn't happen overnight. It took time for the people at the company to even know I was out there and working with clients.

But over time, they kept hearing about the work I've done. They hear about the happy clients and the global communities. And then they took notice. If what you do does not revolve around a specific software, partnering with someone in your industry, that doesn't do exactly what you do, is a good thing for business.

If it's someone you trust, you can refer them, and they can refer you. You don't need to "keep count." Just know that they are out there talking about you. It's always a good idea to connect occasionally to make sure you are both still on the same page.

In order to get what you want out of networking and referrals, keep talking about what you do and be clear on who you serve. Don't try to be everything to everyone. I know this. When I started out, I helped women with technology. "What kind of technology?" you might ask. And it would be a valid question. I thought all the technologies; I can help with anything that's technical, but really that was too broad an area. When I focused on working with entrepreneurs to strategize and implement an online community platform, called Mighty Networks, everything started to fall into place. It was clear what I did and who I served.

So, keep in mind, to spread the word about what you do does not have to be hard. Or feel salesy. People are out there looking for someone who does exactly what you do. Keep the word spreading so when something comes up, people think of you. When you have your own business, people will support you. They want to see you succeed. And feel free to send me a message to let me know what you've done and how it felt.

KEY TAKEAWAYS

- Be part of a networking group that has members who support each other.
- Tell everyone what you do and ask them to recommend you if they know of someone who is an ideal client.
- Be clear on what it is you do.

VISIBILITY

12

PUBLIC SPEAKING TO ATTRACT CLIENTS

MICHA GOEBIG

The first time I thought I'd die of embarrassment on stage was in eighth grade. Having coauthored the year's school play—a British-themed cozy mystery—I cast myself as the young heiress. My vision for the character included stiletto heels, courtesy of my mom. Despite diligent daily walking training—you guessed it—I fell. On premiere day. Right there, in the center of the stage.

I was mortified. For a moment, I contemplated running off, but with all the teachers, students, and parents in attendance, hiding was not an option.

So, I got back into character—a spoiled brat—and snapped at my friend Alex, who played the butler, to help me up and then go fire whoever hadn't straightened the (nonexistent) carpet. The audience laughed, and I got an excellent lesson on how to survive a public nightmare: just pick yourself up (or find a butler to do it) and keep going.

After the play was over, I was surprised to find that neither my peers nor the adults commented much on the incident. Instead, they talked about who they liked, what made them laugh, and what caught them off guard. I believe to this day that I would have never set foot on a stage again—never mind making public speaking my main marketing tool—if they had responded differently.

In hindsight, it was my middle school proof point of Maya Angelou's famous quote: *"People will forget what you said, people will forget what you did, but people will never forget how you made them feel."*

With our school play, we made people feel happy and well-entertained.

And every single time you take to the stage to give a talk, it's in your hands to make people feel special, too.

Today, between 50 to 75 percent of my clients come from speaking gigs. You could argue that I attract clients this way because I enjoy giving talks and have a lot of practice. But I strongly believe that anybody can utilize this tool to market their business successfully.

There are a few lessons I have learned over the decades that can make public speaking as a marketing tool easier and more effective for you, too.

Why Public Speaking is Powerful in Marketing

Trust sells, and trusting the human being right in front of you is a lot easier than trusting a website. When people get a real taste of you, including your body language, they tend to "fill in the blanks" of what working with you will be like.

Plus, at in-person events or smaller virtual gatherings, people can chat with you afterward. I have found that the people who are willing to stick around until the end, almost always want to know how to work with me. If you have your calendar link ready, you can book a follow-up conversation right then and there. This is how I have created a lot of business for myself.

If this inspired you to try speaking to grow your business, let's look at some lessons I've learned:

Define Your Comfort Zone

You might argue that public speaking is not in your comfort zone at all. I dare you to cast your net a little wider: There is probably a setting in which you are comfortable taking the lead. It may be a group of friends, your family, a PTA meeting, or a professional group. Leave aside for now that these may not be your ideal clients and think instead about what makes you feel safe to speak up in this setting: Is it a large or small group? Are people interested in the same topic or do they want to learn something new? Do you know them, or are they strangers? Do you meet in person or virtually?

I realized that I am least worried about messing up when speaking to a large group of strangers in person. So, I gave my first talks at chamber of commerce events, rotary clubs, and educational institutions. Later, I moved on to industry conferences where I can meet many ideal clients.

If feeling safe for you involves small groups of people you know, online meetings, or staying off camera, start there: with your peers, with Lives on social media, or with podcasts. There's plenty of time to move beyond your comfort zone—and raise the number of ideal clients in your audience—as you build your experience and confidence.

Find Your Style

I grew up with a dad who was an amazing and very experienced orator. I learned most of what I know about public speaking at home. This turned out to be a mixed blessing: how do you think people respond to a young woman on stage, speaking and moving like a male senior executive? Not well.

It took me a while to figure out why I had such a hard time connecting with people. What did the trick was a video recording an organizer shared with me. I was shocked: That person on stage looked like my robot clone, someone who had donned an invisible gravitas cape that swallowed all high energy and personable demeanor. I was professional but inauthentic. That's why I couldn't reach people or, to circle back to the Maya Angelou quote, make them feel something, *anything*.

This only happened when I allowed myself to ditch a lot of what I had learned and show up as myself. And when do you know that you have truly reached people? Simple: when they start sharing their own stories with you.

Make Your Talk About Your Audience

Obviously, your marketing talk will be about a topic relevant to your ideal client. But *how* do they want to hear about it? What's their current situation? And how do you want them to *feel*?

If your service caters to a personal transformation, sharing your own story is the perfect starting point. When I speak on self-doubt and confidence, I want my audience to feel empowered and excited about change. I may tell them how I spent over three years *not* doing what I love after I moved to the US because I didn't believe I had anything to offer to people in my new home country. What's your story?

I don't always share a personal story, though. Sometimes, a client story might be a better choice or, if you are in B2B, a vivid best-case/worst-case scenario that can rope people in. Adding interaction with your audience can also create a strong sense of belonging, one of the most positive and powerful feelings there is.

I also believe in providing massive value to my audience. Yes, I want people to pay to work with me. I believe, however, only a fraction is willing to do that. If I give everybody effective, tangible steps to take on their own, they are much more likely to turn to me when they are ready to buy. Or they'll sing your praises to others: when I did several free talks at a coding school, I did not win clients from the student body, but I was contacted by and worked with several managers and companies the students interned with.

Nail Your Call to Action

It's perfectly fine to mention work you do for clients during your talk, but don't make it one big sales pitch. That's what your call to action (CTA) at the end is for. Find out beforehand what is acceptable promotion at the event you're speaking at and adhere to the rules.

No worries if there's a no-pitch rule! You can still speak to people after your talk or invite them to hop on a call with you. This is what I usually do anyway, and it works like a charm, especially if you come prepared. For me, a QR code is the best way to get people on my mailing list, a program page, or a call. Taking a picture or screenshot is a gazillion times easier than remembering the spelling of my name (and email address).

Also, if it's an unpaid talk (which is likely), you can usually ask organizers for the email addresses of everybody who registered to send a follow-up. The data is clear: people who heard you present recently are more likely to open your email because they have started to connect with and trust you.

Finally, I'd like to mention that while I haven't taken a public tumble in a long time, I am still nervous every time I step up to the microphone. But I have learned that a bit of stage fright is part of the package and at least as much excitement as it is fear—excitement about making an impact and making new connections. And that's what heart-centered marketing is all about!

KEY TAKEAWAYS:

- Public speaking offers a chance for face-to-face interaction, enabling trust-building and filling in the gaps that written (social media) content might leave.
- In-person events and virtual meetings allow for follow-up conversations, offering an opportunity to continue discussions and establish relationships.

- Begin by speaking in settings where you feel comfortable, gradually expanding to reach more and more ideal clients.
- Embrace your unique speaking style and authenticity to make genuine connections and create memorable experiences for your audience.
- Design your talks around your audience's needs and emotions, sharing relatable stories or scenarios that resonate and provide value.

13

CREATING PRESENTATIONS THAT STICK

HANNE BRØTER

Before 2009, I had no experience creating presentations. That year, my employer asked me to create and deliver a scholarship report for a travel grant I had received. I suspected a printed report would only see the inside of someone's bottom drawer. My message was way too important for such a destiny. Determined to share my passion, I decided to create the report in the form of a presentation. I embarked on the preparations with gusto!

From there, a concept gradually emerged that I have used ever since when creating my presentations and helping others create theirs.

Three keywords are the basis of my concept: VERBAL, VISUAL, and VIRTUOSO. These are easy to remember, and a good presentation includes all three components. Here's where the phrase "practice makes perfect" comes in—to improve on each component, you must practice them with intention.

But before we address these essential parts, let's talk about THE BIG MISTAKE people make when creating presentations; if you take anything away from this chapter, this should be it: humans are genetically disposed to receive information visually and auditorily, but not simultaneously. Listening and reading are mutually exclusive and cannot be executed at the same time. If you don't believe me, try speaking to someone reading a text message!

Therefore, the most awful thing you can do is create text-heavy slides (which are really "documents," not slides). That doesn't mean you cannot communicate verbally in your presentation. You should talk to your audience, but don't ask them to read something while you are speaking (most of us learned to read at an early age and don't need you to read to us).

Have you seen speakers in front of their slides packed with text, bullet points, and incomprehensible graphs who read their slides out loud? That is not a presentation. Your audience may read faster than you and get impatient, waiting for you to get to the next slide.

This condition, known as *Death by PowerPoint*, threatens to silence a medium representing one of the most powerful and personal ways of influencing others. If your communication does not require your delivery to be perceived and accepted, trust that most people can read, cancel the presentation, and issue an email or a PDF.

Effective communication is a crucial part of marketing. You must know how to write, design, and deliver an excellent presentation to share your ideas with others.

Personal, one-to-one contact is the smallest interface between a brand and an individual. In this interface, your brand emerges as a feeling inside the mind and heart of the recipient. A presentation can be such an interface if created right.

Let's look at the three keywords that will help us do that:

VERBAL

Verbal in this context means the written content of your presentation, getting clear about what you are going to convey and why. Here are questions you'll want to answer in your planning phase and beyond.

- **Who will you be talking to?** Who is your audience? Are they your avatars or not? What is their knowledge of your topic? Why have they come to the meeting or presentation? Find out as much as you can about them. Knowing more, how can you best reach them? Address them where they are and imagine having a personal talk with them.

- **How can you help them?** Do they have a problem, and how can you help them solve it? People do not attend your presentation because of you but because they want to get help, solve a problem, or receive some ideas and inspiration on some issue in their lives or businesses that require them to act. What do you want them to do after they have heard and seen your presentation? What is your call to action? People who leave your presentation should feel energized with new ideas on what to do next.

- **What (exactly) are you going to cover and convey?** This question may be a little provoking to you. You are the expert in your field. You know what to say but think about the scope and length of your talk. You cannot put all your expertise into a single presentation. How much will this particular audience be able to take in? What is the agreed length of your talk? When you know your topic, audience, the scope, and the length of your talk, you are ready to start the real idea work.

- **Create ideas, not slides.** I am afraid firing up PowerPoint won't give any ideas. Instead, work with sticky notes, pencil, paper, or a whiteboard. Pour out your ideas (without criticism)—one idea per sticky note—with simple words that contain their essence. Put all the sticky notes on a wall, a window, or a whiteboard, trashing those unsuitable for this particular context. Write out the actual words for each idea/sticky note. With your verbal

communication found, how are you going to present it?
Go through the wording for each idea and decide how
you will present them visually. Stick to keywords—the
most vital points—written out in single sentences or short,
bulleted lists, combined with illustrations to give your
audience the proper associations. Aim for images that
clarify and strengthen your points, in addition to
deepening the understanding of your audience as they
listen to your spoken words.

Be aware: This is a crucial point: if all those words are added to the slides, you won't be creating a visual story; instead, you will have created a document. We cannot read and listen simultaneously, but our eyes and minds can take in the meaning of an accurate illustration while listening.

VISUAL

The critical task at this point is finding the right images. Try going "behind" your words and think of what will illustrate and enhance their meaning. Metaphors are a great place to look; these are visual images or symbols for words that are difficult to visualize literally. How many visual metaphors can you come up with for the word *change*?

Search for images that suit the jargon of your target audience and visualizations that can release the right idea in your audience's mind. Always start with the words when finding images.

Make your presentation visually appealing by paying attention to graphic design and typography basics. The most vital graphic design principles involved in presentation design are:

- **Contrast** – Make sure your text has sufficient contrast to the background to be readable.

- **Font types and sizes** – Use readable fonts in sizes that are comfortable to read.
- **Proximity and white space** – Use proximity and white space to show structure in your content.
- **Margins** – Let your graphic elements have space to breathe.
- **Layouts** – Using a grid is very helpful for keeping a consistent form throughout the slides.
- **Chapter dividers** with consistent appearances help the audience to keep track of the progress of the presentation.
- **Branding** – If you have a visual profile guide for your business, use your brand colors, fonts, and image style in your presentations. In any case, use color consistently.

Read books, articles, or blogs that will inform you on graphic design principles, or ask a graphic designer to advise you.

It is tempting to think that a template from your presentation software will take care of the graphic design part of creating a presentation. However, your content wasn't in mind during the template creation. **Content and structure precede form.**

VIRTUOSO

You should feel like a virtuoso when presenting, which is only possible if you know your material front-to-back and have rehearsed; this takes time.

Your presentation's effect on your audience depends on how much time you have used in preparations. What most people choose to skip is practicing the delivery. Like in sports, practice makes perfect. To be a virtuoso, you must practice the performance. Don't leave yourself wondering what will appear on the next slide.

Except when actively interacting with the material on the slide (live presentations), you should never turn your back on your audience; knowing what you will say and when, your attention should always be on your listeners, not your slides. *They are not your cheat sheets* but your visual means to present your message to your audience. Your PowerPoint document should be nearly meaningless without you because the core and crucial information *should come out of your mouth.*

Be a passionate, confident, happy, and inspired expert! Then, you will also have an impact. Nothing is as powerful as the live deliverance of a message in a personal and engaging way. If, after practicing, you find it hard to perform the vocal part of your presentation with confidence, seek help from a speaking coach.

However, your passion increases your courage, ability, and eagerness to speak. If you have a conviction you want to share, you will not be easy to stop. People don't remember virtuosos for their own sake but for what they gave to their audiences. Find the passion in your life, and you will also find resonance.

KEY TAKEAWAYS

- Create slides that are more visual and less "verbal," using them as cues for what you want to share.
- Use metaphors for your slide's visuals—and don't be afraid to have fun.
- Practice your delivery (Ninja tip: video record yourself delivering your presentation to ensure your delivery is on point).

14

THE ONLY BUSINESS CARD YOU'LL EVER NEED

DEBORAH KEVIN

As an entrepreneur, you've likely heard that you ought to consider writing a book. Your peers may have done so, and you might be thinking about it. But something always stops you. Maybe you don't have the time to write. Maybe it's the fraud factor. Maybe you think you have nothing original to say.

To these excuses, I say, "Nope. Each of these come down to one thing: FEAR." Fear of failure. Fear of being visible. Fear of getting something wrong. Fear of [fill-in-the-blank].

And having these fears is perfectly normal. They stop you from starting your book. They show up as you're writing that sh*tty first draft (SFD). They reappear when you finally send your book off to an editor—oh! those fears get quite loud then! They arrive as you prepare to launch your book, whispering in your ear, "Who are you to claim this wisdom? Someone will disagree with what you're saying. And just wait until your family gets hold of your book."

I promise you these things will happen. But you have a choice. Don't allow fear to rule your life. Instead, stand bravely in the face of fear and say, "Not today; I've got this."

Once you've addressed your fear, and chosen to lean into and move through it, you might think, "Now what? I'm not a writer. Where do I even begin?" These are excellent questions to which I have a fantastic response: begin with the end in mind.

Begin with the End in Mind

This advice may sound a little backward, but it's actually the best place to start. Ask yourself these questions:

- What's my goal in writing a book?
- Am I looking to inspire other women to follow in my footsteps or blaze new trails?
- Do I want to nab speaking gigs?
- What kind of book do I want to write? Is it nonfiction or fiction? A thought-leader book or a novel? A memoir or a children's book?
- Where do I want my book to appear?
- What would sharing my wisdom and experience feel like?

I suggest journaling your responses to these important and foundational questions before putting pen to paper. Once you're clear on your WHY, figure out who your book is for.

Your Ideal Reader

As a business owner, you've likely created an ideal client avatar (if you haven't, I highly recommend that you grab *Your First Year*, read Jill Celeste's chapter, and create one immediately!). If your book is a nonfiction thought-leader or lead generation book, then your ideal client avatar is your ideal reader. Easy peasy!

Knowing WHO your ideal reader is will be important for your writing process but also your marketing process. Reflect on the one person who will love your book as much as you do. Believe it or not, narrowing down your ideal reader to one person will make your writing and marketing roles easier. And, contrary to beliefs, it won't limit who actually buys or benefits from your wisdom! Dig into your ideal reader demo and psychographics by answering these questions:

- Age and gender of your ideal reader, if appropriate.
- Describe your ideal reader's personality.
- What keeps your ideal reader up at night?
- Where does she shop (groceries, clothing, gifts, etc.)?
- What kind of car does she drive?
- Where does she hang out in real life and on social media?
- What social issues are important to her?
- What kinds of books does your ideal reader enjoy?

My last bit of advice regarding your ideal reader is to give her a name! Here's why: have you ever tried to write a letter to "everyone?" It's nearly impossible to make your message clear. But write a letter to your BFF, and the words simply flow. Don't believe me? Give it a try.

Getting Started

This is the trickiest part, honestly. What do you want to say—and to whom do you want to say it? Because this is a business marketing book, focus on how to tease out your story and showcase your expertise in a thought-leader book. First, begin with a timeline. On a piece of paper, draw a line that will represent your entrepreneurial journey. What was the catalyst that got you started? Jot that info on the timeline. What were the defining moments along the way? Mark these on the timeline. The ask yourself these questions:

- What was your big WHY in choosing the entrepreneurial life? Has this changed?
- What mistakes did I make?

- What did I learn about myself and my business by making those mistakes?
- Where did I struggle most and how did I grow through those struggles?
- Who helped me? Why was their support, encouragement, and/or direction important?
- To someone reading your story, what should be her biggest takeaway?

Try to pinpoint ten places on your timeline. These easily represent ten chapters of your book and will become your working outline. Jot down notes about any stories you can incorporate into each chapter.

Writing Tips

Once you have your working outline, set aside fifteen to thirty minutes every day to write. Building the habit of writing is crucial to your success. Select a chapter on which to focus and write away. Here are a few ninja tricks to help you write in flow and avoid the proverbial writer's block:

- NEVER edit as you write. Doing so completely blocks creativity as your left brain will dominate the experience.
- If you're not "feeling" the chapter or section you selected (or it feels like you're pulling teeth to get words on paper), move quickly to another chapter or section. Permission slip: you don't have to write your book in a linear fashion!
- Finish each writing session by leaving one sentence incomplete. This tip comes directly from Stephen King's *On Writing*. Think of your work as a knitting project— stopping in the middle gives you an excellent place to begin tomorrow.

You've got an important story and message to share, just keep going until you get that SFD written.

How Much is Enough?

One question I hear time and again is, "How long should my book be?" They envision a 400-page tome to indicate their brilliance—and are immediately stymied. Here's a truth: attention spans are significantly shorter today than in the past. Nonfiction books that are between 100-150 pages are statistically more likely to be read in their entirety versus longer books. This means shorter rules—and has the greatest likelihood of creating the kind of impact you want it to create.

Which leads to the obvious question: how many words create a published book within 150 pages of a 6x9 book? If you guessed 30,000 words, give yourself a gold star! Anyone can successfully write a book that's 30,000 words—especially YOU.

What's Next?

You've written and self-edited your book. Now it's time to hand it off to an editor, but not any ol' editor. You're looking for a copy editor first, who will correct any grammatical or punctuation mistakes and ensure you've been consistent in naming conventions and style (e.g., point of view, tense, etc.). The right editor will polish your words to make them sparkle on the page, further increasing your confidence and showcasing your wisdom.

Once the book has been professionally edited, it will be laid out for publication and the cover will be designed. If you've decided to self-publish, you'll have to purchase your own ISBNs and invest in a professionally designed cover. If you've opted to work with a publisher, traditional or hybrid, you can leave these details to them.

Leverage Your Book

Have you ever met an author? Think about your impression as soon as you learned that they'd been published. You, too, can have that kind of panache. People will look at you in an elevated way. What do you actually do with your book once it's published? Here are a few of my favorites:

- Create a speaking platform to get your message out into the world. This includes podcast interviews, presentations, and speaking from the stage.
- Get readers to your email list by offering something of value for them to sign up. The most fun is flat SWAG (fun fact: SWAG stands for *Stuff We All Get*) like a bookmark or an autographed bookplate.
- Prop your book on a shelf behind you so that people can see it when you're on Zoom or FaceTime.
- Routinely ask readers to post reviews on Amazon.
- Use your book as your business card at in-person networking events. You'll instantly become memorable!

Wrap It Up

There are so many reasons to write a book, not the least of which are to increase your clarity and confidence—all which make you irresistible to your ideal clients. Your story matters. You have wisdom and experience to share. Don't allow fear to stop you from making a positive impact on those you're meant to serve.

KEY TAKEAWAYS

- It's normal to be afraid throughout the entire book writing, publishing, and marketing process. But don't let fear stop you!
- If you're considering writing a book, but aren't sure how to start, begin with the end in mind. This reverse engineering will help you identify what to write about and who you are writing for.
- Make sure you know exactly who your ideal reader is. This will help you write a book that will have the greatest impact on the lives you are transforming.
- Create a writing habit by writing fifteen to thirty minutes a day. Your writing doesn't have to be perfect (in fact, it shouldn't be perfect)!
- Once your book is published, think about how you can use it to find your ideal clients, get speaking gigs, and gain greater visibility.

15

CREATING A THIRTY-DAY SOCIAL MEDIA CONTENT PLAN

JENNIFER NICHOLS

Have you ever started to write social media content and immediately begin doubting your ideas, or maybe your nemesis impostor syndrome sits down with you? Suddenly you lose faith in all your ideas while wondering if your message will resonate with your ideal client.

In our fast-paced world, especially when it comes to social media, entrepreneurs are challenged to make the most out of every effort they put in. This is particularly true when it comes to content creation and a solid marketing strategy. The key is to create a balance between minimizing your effort while maximizing impact and gaining visibility. A simple but strategic content plan can help entrepreneurs find the balance between "what do I say" and "will this work" while creating content with speed and efficiency. Who knows, with this focused approach, you may even have some fun!

Let's dive in and create an outline using a step-by-step approach that feels less like work and more like having a conversation with your marketing team.

Step 1: Setting the Foundation with a Thirty-Day Calendar

The first step in effective content planning is to create a clear framework. Begin by looking at a thirty-day calendar. This will serve as the canvas upon which your content strategy will unfold. Having a visual representation of your plan helps you understand how to distribute your content over time.

When looking at the next thirty days for your business, cross out the days you don't intend to post on social media. For example, I don't typically post content on the weekends so I would x out Saturdays and Sundays. If you are like me, you are already getting excited because on average that is eight to ten days you don't even have to think about!

Step 2: Aligning with Holidays and Industry Relevance

After setting up the basic calendar, the next move is to identify key holidays or events relevant to you, your business, or your industry. Websites like www.nationalcalendarday.com can be precious resources for this. I'm a big coffee drinker so you can be sure that every coffee holiday is celebrated on my social media pages. Sprinkle in fun holidays that speak to you, like Root Beer Float Day or National Be Lazy Day. Jotting down these dates on your calendar allows you to craft your content to perfectly match the theme of these awesome occasions.

Taking this step first helps you relax and de-stress before diving into the intense brain stuff. It is also a fun way to engage your audience because, after all, they need permission to celebrate the fun stuff too. I'm old school, so I write my ideas out on paper first; then when I'm complete, I'll transfer the ideas into a social media calendar.

Step 3: Strategic Event Marketing

If you have upcoming events, workshops, or product launches, start marketing them well in advance. Give yourself at least a six-week runway to build anticipation and generate interest among your audience. Add these marketing milestones to your content calendar to ensure a synchronized and effective promotional effort. People need repetition before adding your events to their calendars.

A huge mindset block you may need to overcome can happen when you see your calendar mapped out and all the sales/marketing pieces mapped for the month. You may begin to convince yourself that you're a smarmy used car salesman. STOP! That is a fear belief, and I don't want you to sit with that when planning your content.

Remember that each one of your followers will not see your content every day; busy lives happen, and the algorithm gremlins are always changing things for us and our followers. Marketers contend that the average person needs to see an ad seven to ten times before taking action. That is for paid promotions, so your organic social media content is barely a drop in the bucket for your followers. Posting about your events or products is necessary for you to reach the people that need to be part of your tribe; a few times a week in different formats is not only OK but also necessary. Today it can be a story, in a few days it can be a video. Remember selling from your heart with good intention is a helpful service for those who have been looking for the answers and support you have!

Step 4: Repurposing Blog Content for Consistency

Consistency is crucial in content marketing and that includes duplicating your message. To make this achievable, leverage your existing blog content especially if it doesn't automatically publish to your social media pages. Take pieces of your longer blog posts and transform them into shorter snippets, infographics, or mini videos. This allows you to maintain a consistent presence on your chosen platforms while encouraging your followers to circle back to your website. The best part is that you are educating your followers while also filling up your content cup for now and later.

Step 5: Tapping into Personal Stories and Quotes

For those remaining days on your calendar, draw from personal experiences and anecdotes related to your business or the pain points you address for your ideal clients. Sharing relatable stories fosters a connection with your audience, making your content more engaging and memorable. Incorporate relevant quotes that resonate with your brand's ethos and message.

This is also a good time to test out new ideas like making a behind-the-scenes video, sharing how and why you started your business, or an interaction you recently had with a customer. Testimonials are great here too but more effective if you can add in the story or message of the interaction that led to the testimonial. For example, imagine you are a florist, and someone comes in with a desperate last-minute request. You have a few moments in your day while waiting for a delivery, so you quickly fulfill this request, and everyone is happy!

Three Bonus Tips

1. When in doubt about what content to create, turn to the internet for guidance. Use tools like Google and Answer the Public to identify questions people are asking about your industry or niche. Craft posts and videos that provide clear answers to these questions. By doing so, you're catering to existing curiosity demand.

2. Time management is crucial. Utilize social media scheduling tools to plan your posts in advance. This

ensures that your content gets published even during busy periods, maintaining a consistent online presence.

3. You've worked hard to plan and create this content, let's not waste it, OK? Remember that some pieces of content have a timeless quality to them. These evergreen pieces remain relevant over time. Keep track of these gems in your content plan, making it easier to recycle them periodically, saving you effort in the long run.

As entrepreneurs, we all wear so many hats. Honestly, we are probably wearing more hats than we need to or should! Creating social media content is a hat you probably didn't expect to wear, but it's a necessary one for your overall marketing strategy. Set time aside in your calendar to follow the steps above, and you'll save time and still be able to gain visibility and share your message.

If the steps seem manageable but you still wonder "what should I say," answer the few questions below. Then keep answering them in a variety of ways each month.

- What do I love most about my job or serving my customers?
- What questions (FAQs) am I asked most often?
- If you are product based, what is your favorite product? Why? And how do you use it?
- Tips that you share with customers are also good for social media.
- Do you have a newsletter? Invite people to sign up.
- Poll your audience about something regarding your business or fun, such as the coffee or tea debate.
- Animals/pets win social media every time. I have an office cat who always gets love on social media when I share about him.
- Ask questions. If you are struggling on what new computer to buy or the color to paint your office, ask for

input. If there is a struggle your clients always have, your fans probably do too. Share it!

KEY TAKEAWAYS

- Having a preplanned social media content strategy with buckets of content including words, inspiration, and ideas will help you feel focused and excited about creating your content without all the stress and doubt.
- Consider holidays, observances, and your events when planning out your content calendar.
- It's okay to repurpose your blog posts, newsletter content, and older social media posts.
- Your audience always loves a good story!
- If you're not sure what to post, ask your audience. You can use a poll or post a question and ask for responses in the comments.

16

#MICDROP

KAMIE LEHMANN

I remember the day I hung my open for business sign, posted it everywhere and told everyone I knew. I remember thinking, "They will come."

Well, guess what? They didn't! Truth is, it doesn't matter how great you are, if you have no one to serve, you don't have a thriving business. So, I searched to find how I could attract my ideal clients. I knew they were out there looking for me, but I didn't know how to find and connect with them in this big world of technology.

That's how I came to sign up for a podcasting launch program in February 2020, trusting the claim that I, too, could have a podcast live within twenty-eight days.

When the pandemic changed the world of marketing and advertising, along with the social media algorithms gone mad and the price of Facebook ads exploding, podcasting became an extremely effective way to attract ideal clients. Before the pandemic, there were 850,000 podcasts in the world; today, there are almost three million. All kinds of entrepreneurs use them to market their business.

On March 2, 2020, eleven days before the world shut down due to the COVID-19 pandemic, I launched the *She's Invincible Podcast,* which ranked number forty-two on iTunes in the challenging business-entrepreneurship category. I interviewed fierce women entrepreneurs, highlighting their zones of genius and sharing free education to inspire, empower, and educate women all over the world.

Today, I help challenged entrepreneurs attract their ideal clients, increase their income, influence, and impact through podcasting and monetization. How can podcasting work for you? Here are a few fundamentals that can help your success in leveraging podcasts as part of your marketing strategy.

Be a Podcast Guest

Guesting on podcasts will help you attract your ideal clients by expanding your exposure beyond your circle. They will help you collaborate with referral partners and allow you to sell to your guests and listeners. You're able to share what you do, the result you give your clients, and turn a loyal listener into a paying client.

It's important to be a guest on shows that align with your energy and brand, shows that share the same ideal target audience without creating competition. Focus on collaboration. There is a myth that the famous shows are the best shows. The truth is the best shows are where your ideal clients hang out. You can find shows on iTunes by searching the category that your work falls under, for example, business, education, marketing, finance, health, etc. If you don't know where your people are, find the people who do what you do, and search for them on podcasts to find out where they hang out. That will help you find the right podcast with the right target audience.

Before you pitch yourself to podcast hosts, create a compelling speaker one-sheet and listen to the podcasts you're thinking of guesting on.

Consistency is Key

If your schedule permits, you want to be a guest on one show per week. Share your unique story—what makes you, you? How did you get where you are? What differentiates you from the others in your field? Tell your story, share your journey, the mistakes you made along the way, and what you've learned from those challenges. Be authentic! Be vulnerable!

On your speaker one-sheet, list three to five strong interview topics where you can communicate your expertise and the pain points that you help your clients solve. Include suggested questions to help your interviewer.

By sharing your journey, you become a magnetic storyteller and help your listeners see their own path to success.

Talk About What Led You to Business Ownership

Think in terms of creating a conversation with your host. Podcasts are not infomercials or sales pitches. You're there to give value and impact the lives of the people listening. Go on every podcast with the mindset that everyone paid you $10,000 to serve them.

Infuse Your Mission and Vision into Your Interview

By clearly articulating your mission and vision, your interview will have a different quality. People will get a real sense for what you're up to. While your competitors talk about tactics, when you share what drives you, you will be a totally different energetic frequency.

Have a Strong Call to Action

For your podcast interviews, it's a best practice to have a dedicated landing page where you can direct listeners with a strong call to action to download something of value to them in exchange for their contact information. Once you have their information on your email list, continue to serve and nurture them to build that know, like, and trust factor.

Share Your Interview Generously

When you are on someone else's podcast, you are being elevated to an expert in your field. So put that podcast out there everywhere—on social media, your website, in your newsletter—so everyone can find it (and you). Use the hashtags that describe what you're talking about, what you do, and who your ideal client is. If your podcast host provides graphics, use them. If they don't, create professional-looking graphics yourself. Put the interview in your social media calendar on a regular, rotating basis.

Be a Podcast Host

If you love being a podcast guest, you might consider launching your own podcast. Having aligned guests on your show can help you acquire and nurture new clients. You want to be known for giving value to people. We want to build trust in the company, service, and product by sharing success stories.

By elevating your brand authority by hosting authority figures, you elevate the perceived value of the show. It's a great way to attract future guests. Other guests will say "yes" appearing on your show if you interview inspiring and informative guests.

Instead of paying thousands of dollars to learn from an expert, invite them to your show and ask them everything that you want to know. Do your research, make a list of everything you would like to learn from them, and invite them to your show to ask those questions.

Monetize Your Podcast

As a podcaster, you can monetize your show in a number of ways, namely through affiliate programs and advertising. When a guest on my show has an affiliate program, I include their link on my website, in the show notes, and mention it on the show. This advice works if you're doing podcast interviews, too. Share your affiliate links so everyone can benefit!

At *She's Invincible Podcast* School, we teach our students how to convert their loyal listeners into paying clients. Imagine if you could impact and influence millions of people, amplify your income, expand your impact and influence with a profit-driven podcast, one that is grounded with purpose and profit. How would your life change?

You can sell advertising space on your podcast. People prefer advertising and marketing through podcasting than any other source. Seventy percent of respondents pay more attention to podcast ads than those on television or their social media feeds. Over 50 percent purchased a product or service after hearing it advertised on a podcast. And when a podcast host invites you into their world as a guest, you have a captive audience. Podcast visibility for your business and brand is paramount.

A new study from Magna and Vox Media shows that podcast hosts emerge as the most influential figures shaping people's actions, beliefs, and life decisions. People form strong attachments to their favorite podcast hosts, turning to them for guidance on both financial and personal decisions. Seventy-five percent out of 2,000 respondents placed the influence of podcasters above social media, creators, and mainstream celebrities. Gillian Follett, general assignment editor for AdAge, wrote that nearly 90 percent of millennial respondents credited podcast hosts with influencing their beliefs.

Podcasting has a ripple effect on entrepreneurs' lives, businesses, finances, and futures. It's the fastest way to elevate yourself, your brand, your business, and to get your name out there. Remember, you may never know a million people, but a million people can know you through podcasting.

When you get the right message in front of the right people at the right time, that's where the magic happens! #micdrop

KEY TAKEAWAYS

- Being a podcast guest strategically positions you to attract ideal clients by expanding your exposure and collaborating with referral partners. Focus on shows that align with your brand and energy, targeting the right audience. Create a compelling one-sheet, listen to potential podcasts, and turn loyal listeners into paying clients.
- Consistency in podcast guesting, ideally one show per week, allows you to share your unique story authentically. Share your journey, mistakes, and lessons, becoming a magnetic storyteller. List strong interview topics on your one-sheet, emphasizing your expertise and how you solve clients' pain points. Be a storyteller that resonates and connects with your audience.
- Beyond exposure, monetize your podcast through affiliate programs and advertising. Leverage the influence of podcast hosts to shape decisions and beliefs. Capitalize on the captive audience when you're a guest and explore the potential of hosting your own podcast. Podcasting is not just a medium; it's a transformative tool to elevate your brand, business, and influence in the entrepreneurial landscape.

SALES

17

AUTHORSHIP AND MARKETING: THERE'S STUFF YOU NEED TO KNOW

SUZANNE TREGENZA MOORE

If you've ever considered writing a book believing that it could be a game-changer in your business and skyrocket others' recognition of you as a leader in your industry, you're right. And you are wrong! Stick with me; in this chapter, I'll explain precisely why.

In 2015, I participated in a mastermind where the leader suggested we write a book aligning with our expertise. I'll cut to the chase and tell you I never got mine done. My "shitty first draft" languished, completely outdated, in files hidden somewhere on the interwebs.

Many of my mastermind colleagues, however, did complete their books. This catalyzed me to start my career as a book launch expert, helping them cross the chasm from self-published author to best-selling author using a system I'd learned, tweaked, and continue to use with many clients today. What I want to share with you isn't about the launch; it's about what comes after.

Since my book-launching career began, I've consistently watched new nonfiction authors struggle. Still, I couldn't articulate their challenges and my solutions until I finally wrote my book (a different one than I'd started earlier). So, I'm an observer and a gal right there with you in the trenches, attempting to make a name for herself as an expert in her field.

Of course, some do right! They launch the book as a bestseller, build an empire, make millions, and retire at forty-five. I'm not one of them, and if you are reading this, I'm guessing you aren't either!

The rest of us mortals experience the "new level, new devil" phenomenon and typically find ourselves learning how to market again after the hoopla of our book launch is over. I equate this time with the feelings I experienced after graduation or the weeks after returning home from my honeymoon.

When we work toward a long-term goal, such as writing, editing, and launching a book, there is a letdown: a feeling of "now what?" and confusion about moving forward. Often accompanying this is one of three challenges I've identified and want to highlight for you.

The first is that of becoming a *Muddled Marketer*. Let me define it for you: this is someone struggling to decide what she is actually marketing. Is it a free gift? Is it her nice, shiny, new book? Or is it her large-ticket offer that brings in the big bucks?

The shift from entrepreneur to authorpreneur wreaks havoc for many on what was previously a clearly defined funnel from "irresistible free offer" to "high-ticket offer." New authors fall into the trap of believing that book sales alone will represent a significant revenue stream to their business and should be the primary focus. Reality is quite different. Think again if you believe you will write a book and retire on the royalties. Very few nonfiction writers do this. Instead, they earn their living by leveraging the credibility gained from writing the book.

Their newly enhanced credibility, when appropriately used, gains them speaking engagements, interview opportunities, and status. Harnessing these into leads for high-ticket payment is, and should be, the goal. Book sales are secondary.

Keep this in mind, and you won't be a *Muddled Marketer*. You'll be a thought leader with clarity around what you are marketing to each audience you address.

The second challenge I tackle with my new authorpreneur clients is that of becoming a *Frustrated Follower*. A *Frustrated Follower* finds a guru, often through a Facebook or Instagram advertisement late at night, who seems to have it all together. The guru shares a system for making money that, if implemented correctly, would solve all the *Frustrated Follower's* problems.

Here's the catch! The guru's solution draws on her strengths, desires, and persona. Suppose the *Frustrated Follower* does not align with those. In that case, she will either fail miserably attempting to implement the "step-by-step foolproof system," or she will execute the system flawlessly only to find herself burned out running a business that she no longer loves.

I can personally attest to this as I've done both. The first, more times than I can count!

Avoiding the perils of becoming a *Frustrated Follower* is critical and challenging, but there are two ingredients that make it possible. First, you must understand yourself as a marketer. You need to learn your strengths, your weaknesses, what drains you, and what gives you energy. Second, you must work with a coach or leader capable of helping you develop a personalized plan for success.

Harnessing your individuality as a strength and leaning into it, rather than attempting to improve upon your weaknesses, is always more likely to yield success. Unfortunately, many authorpreneurs focus on what is not working with the belief that improving upon those strategies or tactics will create positive momentum.

My experience shows me an alternate result. I find authorpreneurs who feel out of their depth and begin to lose confidence in themselves. With their self-esteem trashed, they start to question everything they are doing and feel they need an external source to tell them if each decision is right. Eventually, they struggle to believe they ever knew what they were talking about, and completely forget that they are an expert in their field.

Not a recipe for success!

You are not a cookie cutter. Don't think you can fit your marketing solution into one.

Developing a marketing plan that draws on your strengths, and is effective, is imperative. This doesn't mean there won't be elements you don't love to do. Work is called work for a reason. However, once you've achieved some success, you can pass off the elements you don't like to a team member, and you'll watch your business soar.

The third challenge I seen authorpreneurs experience is that of being *Popular but Profitless*. These outgoing authors usually have robust social media platforms, are regularly asked to speak on others' platforms, and consistently share where they can be found.

The problem is that they don't have systems to capture or follow up with leads. So, instead of garnering revenue from all the effort of speaking at events, being interviewed on podcasts, and interacting all day on social media, they are fretting about how they will pay for their customer relationship management system that barely gets used.

These authors have not created a clear pathway that transitions a potential client from being aware of them, to interested in working together, to being credit card ready. Without this pathway, these authors appear to have flourishing businesses and sell many books but are in a cash crunch.

The *Popular but Profitless* authors require an overhaul of their systems. They need items that are easy *yesses* for their ideal clients—kind of like the chocolate bars in the checkout line at the grocery store. Then they need to take those who've said yes on a journey toward a comprehensive, high-ticket offer solution.

Unfortunately, many send a few follow-up emails and then find themselves so busy with social media, speaking engagements, and podcast interviews that they do no follow-up, thus losing the opportunity for a sale. Breaking this cycle is imperative.

Many new authors find themselves bewildered by the pieces and parts of all three of these challenges. In my experience coaching them, breaking each down is the beginning of unraveling the individualized revenue success system needed.

So, if you are thinking that writing a book may be the game-changer you need in your business, you may be right. However, you'll want to be sure you don't fall into the traps that will make you a *Muddled Marketer*, a *Frustrated Follower*, or *Popular but Profitless*.

KEY TAKEAWAYS

- Be clear on your publishing goals and determine what you can gift your readers to entice them to join your mailing list.
- Embrace your newly enhanced credibility, and leverage it for speaking engagements, interview opportunities, and status. Harnessing these into leads for high-ticket payment is, and should be, the goal. Book sales are secondary.
- Create a clear pathway for your potential clients to find, connect, and purchase from you.

18

HEART-CENTERED SALES FUNNELS

CAROLYN CHOATE

What is a chapter about sales funnels doing in a book about heart-centered marketing? Aren't those things contradictory?

There are a lot of myths around what a sales or marketing funnel actually is. We hear those words all over the internet and think, "Do I need that? What is it? How does it work? Is it just for those loud and obnoxious marketers who only care about numbers? Will it fix my lead generation problems?"

The truth is, if you have an online business, you have funnels already. Funnels are the paths that guide people on the internet and help them discover things. Almost anything can be a funnel but if you haven't designed yours, it probably isn't running very well because it happened by accident and not by loving design.

Perhaps you post on LinkedIn to tell potential clients about what you do. But do you just post when you think of something to say? Do you have a plan and a strategy for what action you want someone reading that post to take next?

With a good funnel plan, you know exactly what actions you should take, and you know where those things lead next. It's all set up already and much of it is automated into a happy little machine bringing you potential clients eager to work with you.

Yes, the funnel is a type of automated machine but it's not a conveyor belt for stamping the word "lead" on people's foreheads. It's not going to be an ATM and, when you expect it to be, you miss out because most people will not be ready to work with you at the first encounter. Instead of expecting instant sales or discovery calls booked, keep bringing new people to the first step of the funnel and have the system in place to keep nurturing the relationship and giving them value. When someone is ready to buy, they'll remember who gave generously and who treated them like a waste of space "freebie seeker."

Too many marketers will tell you that their funnel design is special, and you just have to follow the exact steps they did and the next day you'll have a million dollars. But there is no silver bullet in marketing. What works fantastically for one person doesn't do anything for the next. It's not the system that creates the success, it's designing that system to match your values, your clients, and your business. Your funnel will only work if it is created from your heart.

There is no right funnel for everyone, so the question you need to ask is not, "Is it better to use an e-book to bring people to my email list or a live challenge?" The question you need to ask is, "What format will work best for my ideal future clients?"

Here's how an effective funnel works:

1. First, decide the final destination of the funnel. What is the end product or service or item that you want someone to purchase? Then build out the steps backward from there.
2. Spend some time thinking from the point of view of your perfect ideal client. Imagine them not knowing you or what you do at all, but you know they need you. How might they first find out about you? What will catch their attention? How will they know you have what they need?

What will they need to know, believe, or trust in order to take the next step?

3. Write out each step of that client journey and write down any ways that the step can be done without you having to be there (e.g., a webinar that explains your system that can be watched any time someone signs up, an email sequence that teaches some of the core principles they will need to understand before working with you, etc.).

4. Look at how online tech can be used to connect those dots and bring someone seamlessly through the steps, so they are supported and guided toward the end goal.

5. For each touch point on that path, you create a part of the funnel. And when you finish, you'll have a funnel completely customized to you and your clients that guides them perfectly step-by-step from stranger to client to raving fan.

Even better, because so much of those touch points can be automated, you can free up your time to do the work you love while this little machine goes to work for you on your website. And no one falls through the cracks or gets lost or forgotten. The funnel guides them to the next step each time.

That can be done in an aggressive and pushy manner or it can be done in a gentle and heart-centered manner. For some people, those intense funnels with fifty upsells fit their business and their values perfectly. For most of us they do not.

The key to keeping your funnel heart-centered is to always remember the goal is to connect people with the help they need. It's a disservice to the people who need you if you don't show them how they can work with you and why they should.

The simplest funnel has just three pieces: somewhere to meet people, somewhere to nurture the relationship, and somewhere to make an offer. It can get more complicated than that if it serves your goals and your clients, but it can also stay simple if that works for you.

Another big misconception is that you need special software designed for making funnels. Those products can be nice to have but they are not necessary. The platform is not what makes it a funnel, and you can build your funnel from any online tools, platforms, and software you like. Imagine it like you're making a quilt. Maybe you use a sewing machine because you want to do it quickly or maybe you hand sew it because you want to be able to take the project with you when you're waiting in the car line for your child. The end result is still a quilt. It's not the tool you used that makes it what it is.

There is much more to say about funnels, and I could talk about them all day! I invite you to visit my YouTube channel at Financially Free Author if you're curious to learn more about heart-centered funnels.

KEY TAKEAWAYS

- A funnel can be a gentle and kind path to guide potential clients rather than a pushy marketing machine turning people into numbers.
- There is no one funnel type that everyone should be using; there's only the funnel that's right for you, your business, your values, and your clients. It will only work if it matches your heart's energy.
- To make a heart-centered funnel, create each step with service to your client in mind.

19

ONE CONNECTION LEADS TO ANOTHER

CATHERINE WILLIAMS

Connection and building relationships lay at the heart of any small business—perhaps particularly for those, like me, who are sole traders and wear all the hats of their businesses from CEO to administrator and everything in between.

Connecting with other small business owners is invaluable. How you do that can be a process of trial and error until you find what works for you. There are myriad networking opportunities, including social media—especially Facebook groups and LinkedIn, in-person and virtual networking groups.

When I started my business in 2017, I made some assumptions about where my work would come from, which turned out to be completely erroneous. My contacts with some traditional publishers disappeared, and I realized that I had to start networking and build some relationships quickly or face the prospect of returning to employment.

I knew the market was there for self-publishing people who needed help getting their books ready for publication—but how to find them, connect with them, and ultimately persuade them that I was the perfect person to help took a while to figure out.

Here's the short version of how one connection led to another and another, and another ... to the point where most of my work comes through referrals, recommendations, and repeat work for some clients.

Facebook Groups

I joined some Facebook groups—probably too many—as engaging in more than a few meaningfully is impossible. One of them was a local group for freelancers, which also organized coworking sessions at a coffee shop. At one coworking meeting, I was introduced to Keith, who told me about another FB group for those wanting to use LinkedIn for what is known as "content marketing."

Through this LinkedIn Marketing group, ironically hosted on Facebook, I made more connections and ultimately joined a mastermind program to learn how to quickly make the best use of LinkedIn. I was convinced that LinkedIn was where my audience would hang out online.

In another FB group focused on branding and strategy, I connected with a woman who introduced me to Deborah Kevin. We had a conversation, and Debby became a client. I've been delighted to work on books for some Highlander Press authors.

LinkedIn

I'm not as active as I used to be on LinkedIn. Still, the mastermind program pushed me to regularly post what I hoped would be helpful content for those planning to self-publish a book, with some occasional more personal posts. I connected with editors, proofreaders, other designers, indexers, book coaches, and many others involved in the publishing world. My network multiplied, and my clients started recommending me to people in their networks.

I also had the benefit of connecting with fellow members of my mastermind, which led to taking part in some virtual meetings and getting to know people better without traveling—hold that thought—and joining a selected group at the British Library in London for breakfast meetings.

This particular group became an essential part of my business life. I benefited from some excellent coaching around marketing, business in general, how to stop faffing in particular, and additional personal coaching/neuro-linguistic programming (NLP). My daughter also benefited from career coaching from another member of the group. We're still in contact, and I consider them all to be friends and highly knowledgeable business owners in their particular niches.

Content marketing through LinkedIn and other social media is not a quick way to gain clients and business. When it's done well, however, it makes a big difference to the volume of referrals you can get from connections and people you have already helped. I have found that it's much easier to keep the momentum going once the referral ball starts rolling. Seven years after starting my business, most of my work comes from existing connections either directly or by way of referral.

In-Person Networking

The bottom line is that you must find the right group for you. It's no good to attend formal breakfast meetings if that's not your style. I attended a couple of groups as a guest, but in the early days of my business, I felt really uncomfortable doing my "pitch" to people I'd never met before. I tried a few informal groups—literally turn up to the venue, pay £5, and speak to as many other attendees as possible in the two-hour meeting. I can honestly say I never got a single referral or any kind of useful connection from these meetings, but at least it got me out of the house and speaking to people for a while.

That said, in-person networking groups may work brilliantly for you—a lot depends on the kind of business you run—so they're worth trying out.

Virtual Networking

The pandemic forced us to go online and network via video calls. It made us realize that we don't have to travel to meet people we can do business with and provide and receive referrals. Sure, it's not the same as meeting people in real life. But, especially when time is at a premium, and you work with people who could live anywhere in the world, I've found it to be an excellent way of networking, making more connections, and then being able to both give and receive recommendations.

As with in-person networking, it's worth trying different virtual networking groups to ensure that, if and when you join one or two, it's the right fit for your business regarding values, ethos, and personality.

I joined Virtual Networkers in February 2023 after attending a meeting as a guest. I'd been introduced to it from a Facebook post by one of my existing connections who has been a member since its inception. Suffice to say that I felt an immediate kinship with many of the women attending that meeting.

I was also glad to have found an organization that could help me connect with more people from the USA, Canada, and several European countries involved in the publishing industry, particularly those who help authors to self-publish. After just a few months of membership, I'd had several calls with fellow members and other people I'd been referred to. While I live in the UK, I can work with people from all over the world.

Of course, the nature of the publishing industry is that the lead times can be very long. It takes time for books to be written and edited and thus ready for my services. Nonetheless, I am confident that some of those referrals will yield further opportunities for me to help get more books ready for publication.

The Benefits of Networking

Whichever way you find to network, I hope you will find it as beneficial as I have. The first few connections you make may not lead to immediate work for your business, but the further connections you make as a result of those first few are bound to lead to more work in due time.

To summarize, here's a list of the benefits of networking:

- Career and business advancement—engaging with other professionals from different industries and diverse backgrounds can open up many doors to additional opportunities, leads, and collaborations. For entrepreneurs, networking is a vital means of creating new business leads. Building good relationships with potential clients, suppliers, and investors can lead to profitable ventures and partnerships. After a while, word-of-mouth referrals can significantly boost a business's reputation and customer base.
- Knowledge and learning—engaging in discussions, attending workshops, or participating in online forums helps you learn from others' successes and failures, fostering personal and professional growth. Learning together can also provide inspiration and innovation through exposure to new ideas.
- More confidence and improved communication skills— whether at an in-person or a virtual meeting, giving your thirty-second or longer pitch can be nerve-racking. Attending regular meetings with the same group gives you the chance to practice and learn from others about how to present yourself and your business. And I promise it does get easier with practice—even for those at the introverted end of the spectrum.

- Support—networking gives you access to a support system that can offer guidance, encouragement, and even mentoring. Connection with people who have dealt with similar challenges can give you valuable advice and insights. Mentoring can result in faster personal and professional growth.
- Emotional and social well-being—having a sense of belonging and social connectedness is immensely beneficial to your overall well-being. For those who work from home alone, networking reduces feelings of isolation and can alleviate stress.
- Cultural awareness and diversity—networking with people from various backgrounds gives you a greater understanding of diversity and helps promote a more inclusive society.

Networking opens doors to new opportunities and enriches life through meaningful relationships. We are fortunate to live in an era where technology facilitates global networking. Those who embrace networking will have a definite advantage in our connected world.

KEY TAKEAWAYS

- Try a number of different networking groups to find what works for you and your business.
- There are many benefits to networking aside from increased business opportunities.
- It really does get easier with practice.

20

DRIVING SALES USING PUBLIC RELATIONS

ANDREA PASS

Do you need to be famous and on the red carpet to secure press interviews or features? No. Do you need a large budget to be quoted in digital media? No. Do you have to pay for earned media coverage (also known as editorial placement)? No.

In fact, entrepreneurs and authors just like you have secured targeted opportunities including:

- Television and Radio: local market, national broadcasts, syndicated programming.
- Digital: online magazines, newspapers, blogs.
- Print: magazines (consumer, trade, business), newspapers (local market and national), newsletters.
- Podcasts: consumer and business podcasts which result in new audiences every single day.

Here are a few examples:

- A client who is a therapist for teens wanted to increase her client base. She wrote a book, and I secured ongoing podcast interviews and digital magazine quotes and reviews. The client shared the press coverage on her social media pages, tagging the media outlet. The result: a full calendar of patient bookings and plans for the next book.

- An expert in at-home inventing wrote a book to explain the right way and wrong way to get a consumer product/invention to market. Interviews on television, radio, podcasts, and quotes and features in digital press grew sales of the book, additional clients, and speaking engagements.

- A consumer products company looked to increase sales and get picked up by known retailers. A public relations (PR) campaign reaching targeted buyers resulted in product reviews, contents/giveaways, interviews with the company founder, and more. Due to the tremendous amount of media coverage, the brand was able to secure shelf space in major retail outlets.

- An autistic high school student wrote a nonfiction book. After receiving media training, the youngster was featured on podcast interviews and interviewed for features in key press. Product companies interested in his backstory reached out to consider him as a potential spokesperson.

How Public Relations Differs from Advertising

There is power in PR. But you may be wondering: how is public relations different from advertising?

In advertising, you pay for space or airtime. The public knows you created the content, and you are selling the reader, viewer, or listener. Therefore, the business or consumer absorbing that ad may be skeptical. You are telling them what you want them to know, but they don't know if it's the right purchase for them.

In public relations, you are included in earned media or editorial content. That means the journalist has a responsibility to tell the truth and to secure information that is informative and, oftentimes, educational to share with the audience of that newspaper, magazine, television or radio broadcast, podcast, newsletter or blog. In PR, you are not directly selling. The audience wants to learn from you and wants to hear your expertise.

In advertising, you spend a great deal of money. You know that the ad will run on a specific date in a special media outlet. Oftentimes, the public skips right over the ad.

In public relations, the audience is choosing to engage with the content. They are reading, watching, listening, and engaging with the information you are sharing. Whether you are discussing a book, business, product or nonprofit, you have something important to say, and securing press coverage is a way to present that message to audiences that want to learn it.

Finally, while advertising is fleeting, PR lives on. You can share the media coverage with your social media audience over and over again. The date does not matter if the information is evergreen and not tagged to a specific event. Therefore, public relations gives you more bang for your buck! It is an affordable addition to your marketing plan.

Prepare a Press Pitch to Reach Targeted Media

How do you get the media's attention? While some might immediately say that a press kit is the most important element of PR outreach, that approach is, oftentimes, antiquated.

Start by emailing the journalist, highlighting the who, what, where, when, why, and how. Those covering a specific category should easily grasp what your business does, and why it is relevant and newsworthy to a journalist covering that vertical industry.

Remember to include the "meat"—what makes your story unique and what is your value proposition. A press release is helpful in communicating that message to the media while an easy-to-read pitch effectively notes what you have to say and why the press should be interested in learning more about your company, product or service. Stay focused on the message. Remember: less is more.

For a new product, service, company, or speaker news, be sure to have a fact sheet, and product photography or a biography and head-shot of the speaker. This is helpful in rounding out a potential story.

Stand out! By effectively connecting with the media, you are, in turn, connecting with current and potential customers.

Note: journalists are pressed for time. Prepare three key message points to review prior to any interview. If you are concise and on target, the writer or host will best grasp the topic during your discussion.

While each media contact will want something slightly different, the spokesperson should target responses and keep answers concise. Nothing is "off the record."

Strive for Evergreen Content

If the public has been waiting anxiously for your business, product, or book to launch because you are a well-known business or enter-tainment personality, then a date is relevant. For the rest of us, the launch date simply doesn't matter. It won't impact press or sales. In fact, it is best to stay evergreen when talking about a business, prod-uct, or book so that the editorial content will be evergreen too. That means you can use that media coverage over and over again without it being outdated.

Always plan in advance. Remember that the press is planning the Christmas/Hanukkah season beginning in June of that same year. It takes time for media to express interest and prepare content. For back-to-school, begin in May. For Valentine's Day, start outreach in November. For summer, kick off PR efforts in January/February. Always think ahead.

Each relationship solidified with a journalist can become a lifetime connection resulting in additional interviews or quotes to benefit your business.

Secure All Press Coverage; Highlight on Social Media

Every press placement is an opportunity to use that interview, review, or feature to effectively market a business. An integrated message communicated through social media channels further establishes a service, product, or brand with targeted audiences. Traditional press coverage coupled with social media re-posts allow coverage to live on long after the individual trade show, conference, or meeting.

Therefore, share the interview, feature or quotes when the media placement appears. Then, share it again in six weeks, three months, five months, eight months, etc. The point is that each time you share evergreen content, you are opening that coverage to new audiences.

Finding the Right PR Professional

It's all in the gut. Take the time to speak to a number of PR consultants or firms. Do your homework. Don't believe the person who guarantees you press coverage. No one can guarantee press coverage. Work with someone who is authentic and honest, and explains the process for securing interviews, quotes, and reviews.

Public relations is a process. It doesn't happen overnight. Interviews take time to book, and then, they take time to appear.

Work with a publicist who you connect with. I meet people, and we sometimes don't click; therefore, we won't work well together. Many other times, I meet connections, and we know, right away, that we are the right fit. I communicate effectively, set realistic expectations, secure media coverage, and become an important part of the team.

Public Relations Outreach Timing

Begin PR outreach as soon as you have a website, sales sites, and active social media pages.

If you are an author, it is helpful to have advanced reader copies (ARCs) available in print more than ninety days prior to print date. However, that isn't a requirement. Once a book is ready to be sold, it is ready to be reviewed. Reviewers prefer printed copies of the book rather than e-books as they frequently take notes or flag pages.

All other businesses should start press outreach today. It's never too early or too late to add public relations to your marketing mix!

KEY TAKEAWAYS

- Leverage public relations as part of your marketing strategy.
- Plan in advance: June is the time to pitch December holiday content.
- Have three key messages prepared for your press pitches and interviews.

21

LEVERAGING EVENTS TO ATTRACT MORE CLIENTS

CATHY AGASAR

Planning events has always been part of my life. Shortly after my mother passed last fall, I found a party-planning guide I created for a class assignment in the sixth grade. Mom had helped me organize my thoughts; we had even talked about a business someday. I was ecstatic when I got an A on the project!

Fast forward to my junior year of high school. I was elected area president for a business organization, Future Business Leaders of America, which meant I had to plan a conference. I used what I'd learned putting together that party-planning guide, added to it, and created a successful event. I continued planning and marketing these kinds of events for the next eight years, thinking this would be my future.

As fate would have it, I moved into full-time marketing, earning a certified, professional marketing director designation. I was still planning events, but now they were marketing events that needed to bring a return; in other words, make money. Whether it was a grand reopening or an in-store promotion, I had to justify every aspect of what was being done. Sure, there was a piece of goodwill in every event, but bottom line, each event had goals for a significant return on the investment.

Event marketing is one of the best ways to grow both visibility and business, but it takes time and effort, as well as a plan and a budget, to achieve the desired goals. According to Hubspot, 95 percent of all companies engaging in event marketing believe it makes an impact on their business goals.[3] Experience tells me that's because they planned it that way.

Marketing events come in a variety of sizes, but they all have one thing in common: increase visibility and grow revenue. The most successful marketing events are the ones that are well planned and executed; the key to leveraging such events is all in the planning. I suggest you plan from the goals backward, and then work the plan forward to make it all come together.

Sound confusing? It doesn't have to be. The saying goes, "If you don't know where you're going, how will you know when you've arrived?" The same is true for event marketing. Always start with the end in mind. What results do you want to achieve and how will you know if you reach them? Knowing where you're headed makes the path easier to follow.

In mall marketing, I was told to "throw up a few balloons and call it a party." When I started specializing in mall grand reopenings, I was told to "make the gift bags look fluffy" and it would all be good. I quickly realized it was more than just a party; there's an actual science to the art form to gain the visibility—and the sales—desired.

Event marketing can fall into several categories:

- Launch parties and celebrations.
- Trade show exhibits.
- In-store promotions.
- Community events.
- Online events.
- Seminars and workshops.
- Networking.

- Sponsorships.

As a business owner, you have to decide which of these events makes the most "cents" for your desired growth.

Every single success in event marketing is planned and executed in detail, including the time and money to be allocated. Every dollar spent should present your company or product in the best light possible. Every person involved should know the goals and what part they play in achieving them. Everything, absolutely everything, is geared to the end result, whether it's visibility into new markets, increased client counts, increased revenues, goodwill, publicity, or something else. If you cannot nail down the details, there is no reason to do the event.

In the spring of 2001, I worked for an educational marketing firm, in charge of planning the company's participation in an industry trade show scheduled for that fall. We had never exhibited before, but this year, we wanted to make a bigger splash, so we planned to host a booth. Five months out, I gathered my team to discuss goals, specifics like how much new business we wanted to generate, which companies to target, what to do to secure additional projects from current clients, and what successful projects to showcase to attract new business.

We set the goals, brainstormed how to achieve them, determined the visuals and materials needed, and created the marketing materials to let attendees know we were showing up in a big way. We sent pre-event postcards and emails, called sales prospects to secure appointments during the show, and trained the team to work the booth. We had our hospitality ready to go and were greeted with much enthusiasm upon arrival at the event. We held our scheduled appointments over the next three days, and surpassed our goals: four new clients, seven new projects from existing clients, and an award for one of the marketing campaigns we produced. The opportunity was deemed a success on many levels.

Two years later, I started my own marketing company. My first event was different as I needed to get my name into the community, so I went to a chamber of commerce business card exchange with a friend who had told me there would be a board filled with attendee business cards at the entrance. I arrived and went straight to that board to see who I needed to meet, and quickly identified ten potential clients. For two hours, I made my way around the room, introducing myself. I met those targets, secured seven meetings over the next week, and when it was all said and done, I had four new clients and three considering future projects with me. My marketing consulting business was launched.

For a client's major new product launch, I used the model of planning from the goals backward and designed a flyer that spoke to the target audience, not the developer. The headline resonated with potential buyers so much that the launch sales were double what the first-year goal was!

When I got into the holistic health arena, I needed to gain credibility as I was in a completely new field. Networking events proved to be a wonderful opportunity for me to gain visibility, garner speaking engagements, and educate people on holistic health. I also ran in-house promotions, like customer appreciation days and gift-with-purchase events, all successful ways to build relationships with my clients to keep them coming back.

And that's what it's all about: building relationships with your customers. Event marketing builds those relationships because they make your target audience feel good and happy. Think about it. When clients feel good about your product or service, client counts and revenues increase.

I would love nothing more than to give you a success plan for your next marketing event, but then where's the fun in that? Event marketing should be an opportunity for you to get your creative on—to get you thinking outside the box of the day-to-day business and allow you to dream about the next level and how to get there. So while there is no one-size-fits-all plan for event marketing, I will share with you a few best practices to leverage events to grow your business:

- **Ask questions.** Does the event align with your mission? Will your target audience be in attendance? What one thing would make participation a no-brainer?
- **Set a budget.** It's more than just the exhibit fee. It's the materials, giveaways, door prize, promotional marketing (website, social media, postcards, etc.) for participation, and booth setup that reflects your business. Oh, and don't forget to add labor hours to the budget so you have ample coverage.
- **Create your space.** What will your booth look like? What will you offer to visitors? Will you set up space for meetings? Will you have a game or activity to draw people in? How will you capture names for future marketing?
- **Train your team.** Tell them what you expect, what to wear, what to say, what to do. Role-play and have fun. Set the stage for success.
- **Follow up.** Make sure your plan has a mechanism for follow-up, whether it's phone calls, a mailing, an email, or something else. Follow up, build the relationship, and close the deal.
- **Debrief.** Review what happened. How did your booth look? How many visitors to the booth? What measurements did you have in place and how did you do? What went well? What should be changed in the future? Did you meet your objectives for participation?

Leveraging events to market your business can be money well spent. Plan it before you work it; remember, it's more than a few balloons. See the end results before you begin, and your path to success will be much clearer and firmer.

KEY TAKEAWAYS

- Marketing events can help you gain visibility and increase your revenue.
- You can host your own event or attend another organization's event, such as your local chamber of commerce.
- Make sure your event is all about your customers and building relationships with them. How can your event make them feel special and happy?
- There is no one-size-fits-all way to host an event. Give yourself permission to be creative!
- Make sure you create a way to follow up with your event participants. Keep the conversation going!

ENDNOTES

[1] Advocacy, Office of. "Facts about Small Business: Women Ownership Statistics." SBA's Office of Advocacy, March 21, 2023. https://advocacy.sba.gov/2023/03/21/facts-about-small-business-women-ownership-statistics/.

[2] "Referral Marketing: Definition, Benefits and Strategies - Indeed." Indeed.com. Accessed December 18, 2023. https://www.indeed.com/career-advice/career-development/referral-marketing.

[3] Decker, Allie. "The Ultimate Guide to Event Marketing." HubSpot Blog, February 22, 2022. https://blog.hubspot.com/marketing/event-marketing.

ABOUT OUR AUTHORS

Cathy Agasar, author of *The Gift of Loss,* **i**s a national board-certified colon hydrotherapist and I-ACT certified colon hydrotherapy instructor. Her own wellness journey brought her to a new level of understanding of the mind/body/spirit connection and a passion to educate the public about gut health and cleansing. Cathy and her husband, Jerry, reside and work in historic Bucks County, Pennsylvania, where they have merged their practices to serve the community together. She is the mother of six children and two cats and enjoys taking long walks, biking, reading, but, most of all, spending time with loved ones. She is a woman of strong faith, grateful every day for the life she is living, and extremely honored to share her journey with all who will listen. Learn more at https://agasarfamilywellcare.com/.

Gabriela Bocanete is an international conference interpreter, trainer, and speaker based near London, UK. She is also a holistic health coach and sound therapy practitioner. She draws on her multi-disciplinary knowledge and expert understanding of stress, its causes and consequences, to offer bespoke interventions that improve brain performance and resilience. Her clients call her live gong meditation sessions the BrainSpa. She is the creator of several online courses and a speaker at professional associations conferences. Learn more at http://gabrielabocanete.com.

Hanne Brøter is a graphic designer, visual branding expert, and teacher of graphic design. Her passion is helping entrepreneurs create and maintain a visual look of their businesses that reflects their brand message in a unique and authentic way. She works with entrepreneurs through her business Your Brand Vision and through Broter School of Design, where she teaches graphic design to non-designers who wants to leverage the power of correct graphic design in their marketing. Learn more at www.YourBrandVision.com.

Jill Celeste, MA, loves Loud Women and loud bassets. That's why you will likely find her teaching marketing and mindset to female entrepreneurs at Celestial University; or facilitating sisterhood and connection through her online networking organization, Virtual Networkers; or hanging out with basset hounds as the co-founder of Tampa Bay Basset Hounds. She's the bestselling author of two books (so far): *Loud Woman* and *That First Client.* She lives near Tampa, Florida, with her husband, two sons, two cats, and a basset hound named Trixie. To learn more about Jill, please visit http://Jill Celeste.com.

Gauri Chawla is a forward-thinking, executive strategist who builds valuable partnerships with global enterprises. She is called on by C-suite leaders to provide insight to advance partnership functions, including building ecosystems that drive value, foster cross-functional engagement, elevate brand awareness, and increase market growth. Gauri's network of alliances spans the globe, and she offers distinctive experience in building and coaching international teams across the Americas, Europe, the Middle East, Africa, and Asia-Pacific. Connect with Gauri on LinkedIn: https://www.linkedin.com/in/gauri-chawla-422425/.

Carolyn Choate is the odd writer who fell in love with marketing. She loves to help authors, coaches, and other online entrepreneurs see marketing in a new light. Marketing yourself is a gift that you give your audience. She has published several romance novels under a pen name and built sales funnels for a wide variety of heart-led entrepreneurs. She has many resources available on her website at https://www.financiallyfreeauthor.com.

Felicia Messina-D'Haiti is a Feng Shui and soul coach/teacher, speaker, award-winning educator, and contributing author of several best-selling books, including *Wall Street Journal* bestseller, *Born to Rise* (2023). Felicia's passion is supporting people in clearing the physical, mental, emotional, and spiritual blockages from their lives using knowledge gained from more than twenty certifications combined with her own journeys of discovery, healing, and transformation. Connect with Felicia at www.feliciadhaiti.com to explore her offerings, including complimentary gifts.

Gail Dixon is a speaker, author, and coach who guides people to hear and express the heart's voice that gives their life meaning and purpose. Gail is recognized as a top expert for speakers, thought leaders, and mission-driven entrepreneurs in creating messages that make an impact. As leader of The Heart's Voice Movement, Gail is committed to creating a future where the heart's voice is the universal language that heals the world. For more information, visit www.heartsvoicemovement.com.

Robin Fitzsimons, founder of Robin's Center for Wellness in Platteville, WI, is an intuitive, spiritual teacher and healer, certified holistic life coach, Usui Reiki master, Akashic record reader, angel reader, certified assertiveness coach, and certified personal trainer. She's passionate about following her divine path and helping people. Robin is a powerful, heart-centered, energy-loving practitioner who can't wait to serve you, whether with an individual session, event, class, or online. She will welcome you with open arms and open hearts. www.wellnesswithrobin.com

Joanne Flynn Black is the founder of launch b4. She encourages entrepreneurs to launch before everything is perfect, to launch *before* they feel 100% ready, to get out in the world. She strategizes and builds online communities on the platform Mighty Networks. When she's not building communities, she's off traveling the world. Her travel memoir, *In Motion: Around the World in Love and Heartbreak* will be released this year with Highlander Press. Learn more at www.launchb4.com.

Stephanie Fritts, founder and chief executive wrangler of Exec Wranglers Virtual Assistant Services, was an executive assistant for more than twenty years and quickly realized that her work eliminating executives' "administrivia" helped them be more impactful. She founded Exec Wranglers to bring this support to small businesses on a remote basis, freeing up these leaders from their day-to-day operations and allowing them to focus on their zone of harmony. Stephanie is also a single mom of one amazing teenage boy and two dogs and an avid live music lover. Learn more at https://www.execwranglers.com/.

Micha Goebig, founder and CEO of GO BIG Coaching & Communications, empowers leaders to achieve career and personal goals while fostering inclusion. Micha is a multi-certified coach, bilingual keynote speaker, published author, and member of Forbes Coaches Council. Her company started in corporate communications for the car industry and continues to support some of Germany's leading companies with global and inclusive messaging. Micha's talks and workshops, including her signature confidence program for women in male-dominated industries, reflect her pragmatic, solution-driven approach. To learn more, visit www.michagoebig.com.

Deborah Kevin (pronounced "KEY-vin"), as the founder and chief inspiration officer of Highlander Press, loves helping change-makers tap into and share their stories of healing and truth with impactful books, including her book entitled *You've Written Your Book. Now What?* Debby, a graduate of Stanford University's Novel Writing program, earned a master's degree in publishing from Western Colorado University. She's trekked the Camino de Santiago, and she lives in Maryland with the love of her life, Rob, her sons, and their puppy Fergus—that is when they're not off discovering the world. Learn more at https://highlanderpressbooks.com.

Kamie Lehmann is the dynamic force behind She's Invincible Podcast School. As a seasoned podcast host, coach, and owner, she's no stranger to success. Her podcast launched at number forty-two in iTunes' Business-Entrepreneurship category, a testament to her marketing prowess. With a knack for humor and creativity, Kamie helps entrepreneurs use podcasting to elevate their brands, turning their shows into micdrop marketing experiences. Kamie Lehmann turns entrepreneurs into podcasting pros, one mic at a time. Learn more at https://kamielehmann.com/.

Stefanie Joy Muscat, a visionary and accomplished consultant, founded Bevara in 2004 after a successful career in nonprofit management. A Nantucket Project Fellow and a *Detroit Free Press* "100 People to Watch This Century" honoree, she leverages her expertise and experience through writing, dynamic speaking engagements, and personal philanthropy. Stefanie's mission: empower nonprofits with sound business choices, catalyzing small yet impactful changes. She also guides entrepreneurs in community engagement, embodying her commitment to transformative action. Visit https://www.bevaraweb.com/ to learn more.

Jennifer Nichols is a social media confidence coach who helps clients create social media content while overcoming their fears, frustrations, and excuses. Her passion is helping people see their potential and overcome whatever is holding them back from achieving it.

Jenn knows first-hand how to build confidence and create successful social media content because she's done it herself—after overcoming her own fear of being in the public eye. Visit http://www.bloomand hustle.com to learn more.

Andrea Pass is the owner of Andrea Pass Public Relations. With an expertise in booking podcast interviews, features/quotes in digital media as well as broadcast and print coverage in national, regional and local media, Andrea works with clients in such categories as consumer products, lifestyle, B2B, education, authors, non-profits and more. Her strength in relationships coupled with her knowledge of the ever-growing press base results in securing top tier, targeted media placements to increase brand awareness, reputation management and sales for established businesses and growing entrepreneurs alike. Connect with her at www.AndreaPassPR.com.

Laura Templeton's heart physically hurt as she watched business owners and professionals in her network struggle to share their brilliance in thirty seconds or less. She had a simple formula that could help so she started teaching. As a global speaker and best-selling author of *30 Second Success: Ditch the pitch and start connecting!*, Laura inspires audiences and clients to dig deep and find the words that connect with the people they are meant to serve. For more information, visit https://30SecondSuccess.com.

Suzanne Tregenza Moore helps non-fiction authors focus on revenue and thought leadership status. Since leaving her six-figure job, Suzanne has employed her MBA in Marketing & Entrepreneurship, along with personal experience from living in the weeds of her business, to support clients with strategy, marketing, technology, delegation, and mindset. Clients describe her as "invaluable" and a "gentle butt-kicker." She is the best-selling author of *Hang on Tight! Learn to Love the Roller Coaster of Entrepreneurship*. Learn more at http://suzannetmoore.com.

Clare Whalley, straight-talking no-nonsense business coach, runs Meta4 Business Coaching, and works with creative and ambitious business owners. Celebrating fifteen years in business, she guides businesses through structured coaching programs that share proven strategies to achieve more profitability and balance. Helping business owners navigate key business growth challenges from making more money without the chaos through to motivating and developing a team, for them to grow a business to be proud of, one that inspires and motivates so they can create more of what they love. Author of *Do.It.Now! A practical Workbook to Make your Business Work Harder, Not You.* Clare expresses apologies to her readers for the Queen's English being changed to the colonies' English. Learn more at https://meta4coaching.co.uk/.

Catherine Williams is the owner of Chapter One Book Design based in the UK. She collaborates with publishers, book coaches, and independent authors to make their books beautifully easy to read and ready for publication. She brings a wealth of experience from her twenty-eight-year career in publishing that may be hard to find with other book interior designers. She has a superb eye for detail and her unflappable nature means print-ready and digital files are always delivered on schedule. Visit https://chapter-one-book-production.co.uk/ to learn more.

COPYRIGHT INFORMATION

ALSO FROM HIGHLANDER PRESS

Anthologies

Your First Year: What I Wish I'd Known

Bestselling Books

Thriving Through Cancer by Kelly Lutman

Burnt Gloveboxes (Vol. 1) by Gina Ramsey

The Selfish Hour by Megan Weisheipl

Intuitive Languages by Nicole Meltzer

What Color Am I? by Sarah Patterson

Smooth Sailing by Cheri Andrews, Esq.

Peace in Passing (2nd Ed.) by Maribeth Decker

Hang on Tight! by Suzanne Tregenza Moore

A Path of Oneness by Ellen Feldman

Loud Woman by Jill Celeste, MA

By a Thread by Nicolette Blanco

Your 4 Truths by Judy Kane

You, Me, and Anxiety series by Dr. Robyn Reu Graham

That First Client (2nd Ed.) by Jill Celeste

You've Written Your Book. Now What? by Deborah Kevin

The Gift of Loss by Cathy Agasar

30 Second Success by Laura Templeton

Children's Titles

Betty Bartholomew and the Vanishing Begonias by Connie Jo Miller

Meatball and Birdie by Elle Fox

Glitter Bird by Angie Bird

Lauren and Val Take a Walk by Lauren Eileen

Forthcoming Titles

Fourteen Stones by Kris Faatz

Smooth Sailing (2nd Ed.) by Cheri Andrews, Esq.

Juris Ex Machina by John W. Maly

The Knowing by Kimberly Patton

Sara Smitherson and the Disappearing Snickerdoodles by Connie Jo Miller

Penelope Parsons and the Missing Pomegranates by Connie Jo Miller

My Body Knows. Do I Know? by Ashley Cournoyer-Smith

Burnt Gloveboxes (Vol. II) by Gina Ramsey

In Motion by Joanne Flynn Black

Story Shifts by Anne Fowler Wade

Free Spirit at Free Safety by Joe Zagorski

Gray Matter by Avery Volz

Ripples on the Church Water by Brynn MacDonald

Finding Grace by Deborah Kevin

ABOUT THE PUBLISHER

Highlander Press, founded in 2019, is a mid-sized publishing company committed to diversity and sharing big ideas thereby changing the world through words.

Highlander Press guides authors from where they are in the writing-editing-publishing process to where they have an impactful book of which they are proud, making a long-time dream come true. Having authored a book improves your confidence, helps create clarity, and ensures that you claim your expertise. Learning how to leverage your author business takes your experience to a whole new level.

What makes Highlander Press unique is that their business model focuses on building strong collaborative relationships with other women-owned businesses, which specialize in some aspect of the publishing industry, such as graphic design, book marketing, book launching, copyrights, and publicity. The mantra "a rising tide lifts all boats" is one they embrace.

facebook.com/highlanderpress

instagram.com/highlanderpress

linkedin.com/highlanderpress

tiktok.com/@highlanderpress

www.ingramcontent.com/pod-product-compliance
Lightning Source LLC
Chambersburg PA
CBHW040924210326
41597CB00030B/5173